Construction Foreman's Safety Handbook

by
George Kennedy

DELMAR
CENGAGE Learning™

Albany • Bonn • Boston • Cincinnati • Detroit • London • Madrid
Melbourne • Mexico City • New York • Pacific Grove • Paris
San Francisco • Singapore • Tokyo • Toronto • Washington

Construction Foreman's Safety Handbook

George S. Kennedy

Publisher:
Robert D. Lynch

Acquisitions Editor:
Mark Huth

Production Manager:
Mary Ellen Black

Art and Design Coordinator:
Michael Prinzo

Library of Congress Control Number: 96-42681

ISBN-13: 978-0-8273-7882-7

ISBN-10: 0-8273-7882-3

Delmar Cengage Learning
5 Maxwell Drive
Clifton Park, NY 12065-2919
USA

Cengage Learning products are represented in Canada by Nelson Education, Ltd.

For your lifelong learning solutions, visit
delmar.cengage.com

Visit our corporate website at **www.cengage.com**

Printed in the United States of America
12 11 10 09

Contents

Preface

ACKNOWLEDGMENTS

The information compiled in this book was created with the field supervisor and manager in mind. The purpose of this book is to provide construction managers with information to help them to effectively manage safety and health activities at job sites.

I would like to thank the staff and membership of the National Utility Contractors Association (NUCA) for their support and contributions to the content of this book. Special thanks to Mike Ancell for his editorial assistance. I would also like to thank Janine Reid, author of "What to Say When the Sky Starts Falling" and the National Safety Council, Itasca, Illinois, for allowing me to use some of their published information. Figure 5.3 has been used by permission of the National Safety Council, Itasca, Illinois.

In addition, I would also like to thank my lovely wife Frances, and my daughters Tara and Heather for their patience, cooperation, and moral support.

George S. Kennedy

INTRODUCTION

Congratulations, you have been promoted. Your new title is foreman, project manager, superintendent, or maybe safety director. Along with the title, you get a small pay raise and lots of added responsibility. You are now responsible for production, quality control, scheduling, and worker safety.

Did anyone tell you that safety management is part of every construction supervisor's job? Did the boss tell you that being responsible for the safety of others is one of the most important jobs you will ever have?

Chances are that if you are reading this book, your boss provided it, and you are already aware of your company's safety activities. As a job-site supervisor, you must know many different things about job-site safety management. This book was written to help you understand the basic components of a safety program and what you can do to prepare yourself for a very important part of your job—accident prevention.

Human Costs

Accidents often result in someone being hurt. When a worker is injured, the worker and his family suffer in many ways. First, if the worker survives the accident, he or she will suffer from the pain associated with the injury. Second, the worker may not be able to work in his or her trade ever again. Third, the worker's family will be faced with a lower income until the worker can return to work.

If a worker is killed on the job, the worker's family will also suffer the grief associated with the loss of a loved one. Children often suffer the most from severe depression and mental problems after the loss of a parent.

One foremen tells a story of how he had to tell the family of a mechanic who was killed when repairing the hydraulics on a bulldozer and the blade dropped. The accident was fatal. This happened on Christmas Eve, and the foreman had to go to the worker's home to tell the family. The worker's home was bustling with the smells and excitement of Christmas. He says he will never forget that day and is reminded of it every year.

The last thing in the world that you want to have to do is tell a worker's family that dad or mom will not be coming home because he or she was killed on the job. Foremen and supervisors who have done so say it was one of the worst experiences of their lives, especially if they knew they could have prevented the accident.

A Safe Job Is a Productive Job

Safety, quality control, and efficient production are necessary to any company's success. Unless jobs are kept safe, the cost of doing business in the future will increase. Accidents remove skilled workers from the work force, break up productive crews, and increase insurance costs.

Safety is a vital part of construction management. Every construction manager, foreman, and supervisor must recognize that a safe job is a productive job. Safe workers are productive workers, and work quality improves when workers know they are working at a job site where hazards have been controlled or eliminated.

Today, foremen must wear many hats, and perhaps the most important is that of job-site safety coordinator. Foremen are also responsible for production costs, including lost time, damaged materials, equipment down time, scheduling, and so on. Managers and foremen who believe safety activities cost the company money are mistaken. Companies that implement safety programs have fewer accidents, which results in lower operational costs, lower insurance rates, greater bonding power, improved employee morale, increased productivity, and enhanced public image. Unless safety is promoted and properly managed, production costs can increase dramatically.

No task is so important that a worker's life should be risked to complete the job. Workers should never be placed in situations where they feel that their job depends on taking chances.

As a management representative, you are responsible for worker safety at your job site, whether you manage a large job with many workers or a small job with only a few workers. When promoting the company's safety program, your approach to safety and the decisions you make will set the tone for safety at your job site.

Although construction work is dangerous work, the dangers can be controlled or eliminated if managers, foremen, and supervisors take time to implement the job-site safety program. Preventing accidents will protect workers, save the company money, reduce waste, and prevent costly lawsuits. Planning for safety protects workers from injuries and illness, while saving every construction company's most valuable resource—skilled workers.

Unit 1 – Why Is Safety Important to a Company?

There is only one way to do a job: the right way, which is the safe way. Hazards exist at every job site, and they must be controlled if the job is going to be completed the right way. Therefore, safety should not be treated separately; it should be planned and integrated into production activities.

SAFETY-RELATED RESPONSIBILITIES

The frontline manager controls the activities of the workers. Workers look for direction and follow the manager's lead. Actions speak louder than words, so unless the manager demonstrates a genuine interest in preventing accidents, workers will continue to take chances, use defective tools, violate rules, and perform unsafe acts.

As the leader of a group, the manager must follow all company safety policies and procedures. For example, if the manager does not wear a hard hat, safety glasses, and other personal protective equipment, workers will not wear their equipment or take safety seriously.

Safety is everyone's responsibility. However, at the construction job site, the frontline manager is responsible for ensuring that the company's safety program is set in motion. Unless line supervisors take an active interest in safety, workers will be injured or possibly killed during the performance of their work.

This may be hard to believe, but the frontline manager is the key to the success of a company's job-site safety program. Take it seriously, and give safety the consideration it deserves.

ACCIDENT PREVENTION

Obviously, there are many good reasons for preventing accidents. The most important reason is to protect workers and the public from being injured or killed at the job site. Another reason is financial benefits, such as lower insurance rates, fewer liabilities, fewer Occupational Safety and Health Administration (OSHA) penalties, less damage to equipment and materials, fewer schedule interruptions, and preventing temporary or permanent loss of good workers.

Many owners and prospective customers look closely at a contractor's OSHA logs and safety experience rates before making a contract. These safety measures often reflect a company's level of efficiency, production, and quality control. Owners are also concerned about the liability that could result from accidents on their property.

ACCIDENT COSTS

The cost of accidents is much higher than most people think. What people forget to consider is that insurance rates increase as the number of reported claims increase. The uninsured (indirect) costs of accidents are much higher than the insured (direct) costs.

Direct and Indirect Costs of Accidents

Direct (insured) cost of accidents include:

- medical expenses
- worker's compensation

Indirect (uninsured) costs of accidents include:

- loss of productivity
- manager's time
- clerical time
- worker replacement
- training
- overtime
- damaged material
- equipment damage and downtime
- cleanup and repair
- disrupted schedules
- contract penalties

- third-party liability claims
- insurance premium increases
- property damage

Reducing these costs increases a company's profitability and lowers overhead expenses. Preventing injuries and lowering overhead costs can lead to lower bids and more work for the company. Lower overhead can also result in more jobs.

Accident-free work sites not only benefit each contractor, they also benefit the entire construction industry. Accidents at a job site directly impact a company's insurance rates and add to the losses of the construction industry, which influence insurance rates for all contractors.

Experience Modification Rate

The experience modification rate (EMR) is used by the insurance industry to collect higher workers' compensation premiums from companies with higher-than-average losses and to reward companies with lower-than-average losses. EMR is based on the losses of a three-year period that excludes the prior year and the current year. For example, the EMR for 1996 would be based on workers' compensation losses for 1992, 1993, and 1994. A company's EMR is updated each calendar year. If a company can replace a year of high losses with a year of lower losses, the company's EMR will be reduced and premiums will be lower.

An EMR of 1.0 is considered average. If a company has an EMR greater than 1.0, the company pays more for insurance. Companies with an EMR less than 1.0 receive a discount.

Each job classification, such as carpenters, masons, and excavation workers, is assigned standard industrial codes (SIC). A rate per $100 of payroll is established in each state for all classifications.

Workers' compensation premiums may vary from state to state. However, they are generally calculated using the following overly simplified calculation:

$$\text{Payroll} \div \$100 \times \text{WC Rate} \times \text{EMR} = \text{Premium}$$

Example: The workers' compensation rate for sewer construction in New York State as of December 1, 1995, was $12.49 per $100 of payroll. If three contractors with different EMRs were bidding on the same job and each company had an estimated payroll of $100,000, the insurance premium for each company would be as follows:

Company 1 (company just starting out with an EMR = 1.0): $100,000 ÷ $100 × $12.49 × 1.0 = $12,490

Company 2 (company with a good claims record and an EMR of 0.7): $100,000 ÷ $100 × $12.49 × 0.8 = $8,743

Company 3 (company with a poor claims record and an EMR of 1.3): $100,000 ÷ $100 × $12.49 × 1.2 = $16,237

As you can see, company 2 would have a competitive advantage of $7,494 over company 3 when bidding this job. This shows how fewer accidents increases a company's competitive edge.

Liability

With lawsuits so prevalent in this society, contractors must protect themselves. The first step is to have a good liability insurance policy. The second step is to implement an effective safety and health program to protect workers and the public from injury.

Although, most state workers' compensation laws limit workers from filing a lawsuit against their employers, attorneys are finding ways around these laws. For example, if a worker was injured using a defective tool, the injured worker could file a lawsuit against the tool manufacturer, who could file a lawsuit against the employer for not properly maintaining the tool. This is known as third-party liability.

Employees of other contractors at the job site can file a lawsuit against a company if they are injured as a result of the company's activities or while using the company's tools or equipment. For example, if a worker borrows a ladder from a company and falls from that ladder, the company could be held liable. To prevent this type of situation, work areas should be kept safe, unauthorized persons should be restricted from work areas where possible, and tools and equipment should not be lent to other contractors or their employees.

Another source of liability is curious neighbors, children, and pets. Unauthorized people should never be permitted in the work area. Set up fences, barricades, traffic-control devices, warning lines, and so on to keep people and vehicles out of the job site. If someone is injured in the work area, the odds are the company will end up in court and the manager will be called to establish what was done to keep people out and prevent accidents.

Penalties

In addition to increased workers' compensation rates and liability, a company also could be penalized by various government agencies. The most commonly known federal agency is the Occupational Safety and Health Administration (OSHA). OSHA's primary purpose is to ensure that workers are provided with a safe place to work. This is accomplished by establishing safety standards that employers must follow. To ensure compliance with the standards, OSHA compliance officers make inspections and issue citations for violations. In states where state OSHA plans have been established, state compliance officers are given similar powers to inspect work sites.

When citations are issued, penalties often follow. The penalty depends on several variables and can be as much as $70,000 for a willful violation. The penalty for a serious violation, which is a violation that could result in a serious injury or illness, can be as much as $7,000 per violation. A willful violation (i.e., a violation where the employer or job site manager willingly, knowingly, and intentionally disregarded a standard) could result in a penalty of up to $70,000 per violation.

Although, OSHA is required to take company size into account when issuing penalties, companies are often required to pay very expensive penalties. The cost of penalties comes directly from company profits; therefore, compliance with the standards is necessary. More important than any cost, compliance with the standards can prevent workers from being injured.

Other government agencies are also looking out for the safety of workers and the general public. These agencies include the Environmental Protection Agency (EPA), Department of Transportation (DOT), and Department of Environmental Conservation (DEC), as well as various state and local government agencies. Each has its own specific agenda, but in the last few years, cooperation between agencies has increased and they are sharing information. This has resulted in OSHA notifying EPA of environmental hazards that were observed during an OSHA inspection and EPA notifying OSHA of unsafe work sites. Many state and local officials also are cooperating by notifying the appropriate agency when hazards are observed.

PUBLIC RELATIONS

Nobody wants bad publicity. When a serious accident occurs, the press is often the first to know and the story spreads like wild fire. Bad publicity can seriously damage a company's reputation.

Today, many customers verify a company's accident record and experience modification rate before awarding a bid. In addition, negative publicity can cost a company a job. Good customers do not want to hire a company that has a poor safety record or a history of employee injuries. They are concerned because they can be held liable for the actions of the company.

The frontline manager should take the time to prevent accidents. Over the long haul, efforts will be rewarded.

Unit 2 – Accident Causes and Controls

CAUSES OF ACCIDENTS

Accidents do not just happen, they are caused. There are two causes of accidents: unsafe conditions and unsafe acts. An **unsafe condition** is a hazardous physical or environmental condition that could lead to an accident. Examples include a damaged ladder, frayed extension cord, opening in the floor, or chemical vapors. An **unsafe act** is the violation of a commonly accepted safe practice or procedure, such as not wearing personal protective equipment, unauthorized operation of equipment, or failure to report a hazardous condition. Most accidents are caused by a combination of an unsafe condition and an unsafe act.

Workers who do not perform their jobs safely are a hazard to themselves and to their coworkers. A lack of knowledge contributes to the risk of someone performing an unsafe act. Workers who do not know and understand the hazards associated with their work and at their job sites are more likely to be injured than informed workers. Therefore, the manager or supervisor must make sure that workers have been trained and are aware of the hazards they face each day.

Job-site conditions can change rapidly. Potential hazards may need to be pointed out daily or sometimes even more frequently. The manager should heed this rhyme: If in doubt, point it out.

HAZARDS

Hazards exist in many forms. A hazard is a condition that places someone or something at risk. A hazardous condition exists when somebody is in danger of being injured or is exposed to a hazardous environment that could affect their health. The threat of an inanimate object being damaged is also a hazard. Some hazards are obvious, and others are not.

7

Figure 2.1. Common hazards found at construction sites

8

Hazards generally can be separated into two basic categories: physical and environmental. Other types of hazards that are frequently overlooked are the unsafe actions taken by workers and a worker's lack of knowledge.

Hazards at construction job sites are changing constantly as the job progresses. Some specific types of hazards typically found at construction sites include falling on the same level or from an elevation; being struck by falling materials or equipment; getting caught in equipment or between objects; and receiving electrical shock from ungrounded, frayed, or damaged electrical equipment and wiring or from making contact with overhead power lines. OSHA is now especially emphasizing these hazards because they cause 90 percent of all construction fatalities and serious injuries.

How Hazards Interrupt Work

When a serious accident occurs, most workers stop work to see what is happening. The job site comes to a standstill while workers watch the rescue and treatment of the victim. Workers generally stand around until the victim is transported to a medical facility. Following the removal of the victim, workers also spend time on that day and on following days standing around talking about the incident.

Following the accident, the project manager, supervisor, foreman, and the safety director must perform an accident investigation to determine the cause. The accident investigation's purpose is to prevent a similar occurrence in the future. The details of accident investigations are discussed in a later chapter.

As the amount of time wasted accumulates, schedule delays frequently are encountered. If equipment or materials are damaged, additional delays occur.

If the accident is serious enough to warrant an investigation by OSHA, EPA, state or local inspectors, insurance claims representatives, or others, more valuable time is lost. Work also may be interrupted or suspended until the investigation is complete. Some government agencies have the power to shut the job down until the investigation is complete.

Hazard Control

Inspections, audits, training, equipment, and preventive maintenance are some ways to control hazards and prevent accidents. By eliminating the unsafe act and unsafe condition, most accidents can be prevented.

For example, identifying and eliminating defective tools or equipment from the job can lessen the possibility of an accident. When tools or equipment cannot be immediately removed from the job site, they should be tagged clearly "defective." Workers should be instructed not to use tagged tools or equipment.

Employee actions are more difficult to control. Unless the worker knows how to perform safely, he or she probably will make mistakes. Therefore, training each employee to perform specific tasks and to work safely are the first steps in the control process. The next step is to observe the worker in action, to make sure the worker is performing the job as instructed. Any worker that does not follow the established procedures and safety rules should be told to stop work immediately. Never allow the worker to continue, even for a moment.

Figure 2.2
Taking immediate action can prevent an accident.

At this time, the manager must determine why the worker is not working safely and then take action to eliminate or control the hazard before the worker is allowed to continue. If necessary, the worker should be retrained before injuring self or someone else.

When a worker is allowed to continue working in an unsafe manner, the manager's approval is demonstrated. Permitting a worker to perform unsafely sends a message to all workers that the manager is not serious about safety. Allowing workers to perform in an unsafe manner is the start of a unsafe job site.

Unit 3 – Safety Committee

PURPOSE

The primary purpose of safety committee meetings is to exchange information. Safety committees are a forum for identifying hazards and controlling safety-related problems. Supervisors and workers often know more about what actually is taking place at the job site than managers. Workers can provide valuable assistance when developing solutions to safety problems. Because more heads are better than one, committees that include supervisors and workers can suggest methods for eliminating or controlling hazards.

TYPES OF SAFETY COMMITTEES

Management safety committees are made up of managers, foremen, and supervisors. On large job sites with multiple employers, each contractor working at the job site should be required to send a representative to the meetings.

Employee safety committees are made up of workers. Workers are selected from different trades to represent the company's work force. As the eyes and ears of the work force, committee members often can identify problems that easily could go unnoticed. Workers often turn to employee committee members without fear of reprimand. Workers are also more open with fellow workers than they are with management representatives.

Labor-management committees quickly are becoming the preferred type of safety committee. However, there are laws that govern labor-management committees.

Made up of employees and managers representing the companies at the job site, work site committees have proven very effective. OSHA has taken the lead in this area and is encouraging employers to establish labor-management safety committees.

The key to a successful labor-management committee is the establishment of clear objectives and a clear understanding of the committee's purpose. These safety committee meetings should not be used by management or labor as "moan-and-groan sessions." To be successful, they must focus on safety and health issues.

MEMBERS

Managers, supervisors, foremen, and key personnel who are assigned to safety committees should be indoctrinated before attending their first meeting. They should be informed of why they have been selected and what is expected of them. They should understand their responsibilities and how they can help the committee. For example, supervisors and employee representatives can observe the everyday working conditions and communicate with workers in the field. Based on their observations and communication, their input is extremely valuable. If they do not share this information with the committee, it is useless.

HOLDING MEETINGS

The time and place of the safety committee meeting should be established in advance so that all members can plan their schedules. Members should understand that attendance is required, and they should not be excused from attending without a good reason.

Successful meetings depend on an agenda that is well planned and distributed in advance, so committee members have time to collect information and prepare for the meeting. The chairperson should send an agenda to each committee member.

Meetings should be held only when they are necessary. If the meeting is not necessary or if several members of the committee cannot attend, cancel or postpone the meeting. It is important to remember that committee members have other responsibilities, so their time should not be wasted.

13

Safety Committee Meeting
XYZ Construction Company
Main Street Job Site
Thursday, June 5, 199X
10:00 AM to 12:00 noon

I. Introduction
Have everyone introduce themselves if they do not know each other or if you have visitors attending.

II. Purpose of meeting
Tell everyone the reason for the meeting.

III. Monthly accident report
Provide a description of each accident that has occurred since the last meeting. Discuss what could have been done to prevent the accidents. If the problem is not solved, discuss what could be done to prevent another accident.

IV. Job site inspections report
Review the results of the monthly inspections. Report what was done to correct the hazards that were observed.

V. OSHA report
Review the results of any OSHA inspections that may have occurred. Discuss citations that may have been issued at other sites.

VI. Activity report
Review previous action plan items. Verify that the problems or hazards have been corrected.

VII. Develop an action plan
The action plan will be used to identify problems and hazards that need attention. An individual or a task force should be assigned to solve the problem or correct the hazard.

VIII. New business
Other business that requires the committee's attention.

IX. Next meeting
Set the date and time for the next meeting so the committee members can plan to attend the meeting.

X. Adjourn

Figure 3.1
Sample committee agenda

Action

A plan of action should be prepared at each meeting. As problems are identified, actions should be taken to address each problem. Actions might include establishing a method to eliminate or control a specific hazard, or a task force may be assigned to study the problem and to report their findings and suggestions to the committee.

The plan of action should:

- determine what action will be taken;
- determine who will be responsible;
- establish a time frame for completion; and
- follow-up to ensure the action was completed.

If you want your safety committee to be successful, remember that committee members must experience a sense

---

A/P 96-1   Determine why workers were observed standing on the top step of ladders. What can be done to prevent this from happening in the future? *Answer could be as simple as providing taller step ladders or employee training.*
Who: *Personnel assigned to the task*
When: *Date this should be completed*

A/P 96-2   Inspect all electrical tools. *This should be completed by someone who is familiar with tool-related electrical hazards.*
Who: *Personnel assigned to the task*
When: *Date this should be completed*

A/P 96-3   Schedule a trenching and excavation competent person training program. Determine how many people will attend, where it will be held, the date, and the time. This also may require coordinating the schedule with the production schedule, the project manager, the instructor, and other people.
Who: *Personnel assigned to the task*
When: *Date this should be completed*

Figure 3.2
Sample action plan

of accomplishment or they will lose interest and the meetings will be a waste of time and money. Only committee meetings that produce results are worth holding. A good action plan will add to the success of your safety committee.

The Interactive Method

The interactive method is directed at making meetings work. It can be used for almost any type of meeting. As for safety committee meetings, it works very well because it forces members to get involved, be creative, confront problems, and, most of all, participate in solutions. Decisions are made and priorities are set based on committee action, not just the action of a few individuals. The idea is to find win/win solutions that everyone can live with. Involvement leads to acceptance of the solutions.

The interactive method focuses on running task-oriented meetings in which people get together to accomplish a goal— to do work. Task-oriented meetings include a range of work sessions from information sharing and problem solving to planning, evaluating, and decision making. A primary problem with committees is that they often are not focused. By utilizing the interactive method, the committee will focus on one item at a time.

Before taking the first step, committee members must understand the basics of using the interactive method. There are four well-defined roles and responsibilities that collectively form a self-correcting system of checks and balances. All four roles are equally important. Each role contributes to the actions and success of the committee. No one person is in the traditional leadership role, even the chairperson. Everyone is equally responsible for the committee's success or failure. The chairperson does not dominate the committee, unless the committee is at a standstill and immediate action is required. The four key roles are facilitator, recorder, committee member, and manager or chairperson.

The first step is to determine on **what** (problems, topics, or agendas) the committee will focus. This is done through brainstorming sessions focused on important issues or problems.

1. Facilitator

- a neutral servant of the group who does not evaluate or contribute ideas unless requested to do so by the committee,
- helps the group focus its energies on a task by suggesting methods and procedures,
- protects group members from attack,
- prevents anyone from dominating the discussion, and
- makes sure that everyone has an opportunity to participate.

2. Recorder

- a neutral servant of the group who does not evaluate or contribute ideas unless requested to do so by the committee
- responsible for writing down basic ideas on large sheets of paper in front of the committee so that ideas can be preserved and recalled at any time,
- does not edit or reword the ideas, and
- can also be the facilitator.

3. Committee Member

- an active participant,
- keeps the facilitator/recorder in their neutral roles and makes sure that ideas are recorded accurately,
- has control of what happens,
- can make procedural suggestions, overrule the suggestions of the facilitator, and generally determine the course of the meeting, and
- primarily devotes his/her energies to the task.

4. Manager/Chairperson

- an active participant,
- maintains all other powers and responsibilities,
- makes final decisions and has the power to set constraints and regain control if not satisfied by the progress of the meeting
- argues actively for his/her point of view,
- urges the group to accept tasks and deadlines, and
- sets the agenda.

Figure 3.3
The four roles of the interactive method
(*NUCA SAFETY NEWS*, September/October 1993)

Then items should be listed on a large piece of paper where all committee members can see it. At this point, the list should be reviewed and prioritized by the committee. The committee should focus on the items listed.

The second step is to determine **how** (approach, method, or procedure) the items on the list will be accomplished, by focusing on each item, one at a time, in the order of priority. The committee should address each problem the same way they attacked the selection of "what" in the first step. By focusing the energy of all committee members on one item at a time, the committee will be able to provide good ideas and solutions to the problems.

Solutions may not always take place in one session. Many issues and problems may require a task force to research or develop ideas and solutions. The task force is made up of three to four committee members to review a task and make recommendations to the committee. Task force reports must be prepared in writing (brief and to the point) and are sent with the agenda to each committee member before the committee meets. This allows members to contemplate the issue or problem before the meeting, thereby saving time at the actual meeting. Meeting time is valuable and should not be wasted by reports and general discussions. At this point, the committee focuses on each task to formulate a decision or solution. Finally, an action is assigned or taken by the committee.

Unit 4 – Employee Indoctrination

DAY ONE—ALL NEW EMPLOYEES

Before a new employee starts, he or she should receive a new-employee indoctrination. One of the most important topics of this indoctrination is safety and health.

Generally, the new employee will report to the contractor's office or the job site and be informed of the job responsibilities. They will also be required to complete state and federal forms. The employer is required to verify that the employee is a U.S. citizen or has a green card that authorizes employment in the United States. When hired, the employee should be advised of job responsibilities, medical benefits, job supervisors, job-site location, and emergency procedures. Some companies use an indoctrination checklist to ensure that nothing is forgotten.

The safety director, if the company has one, should meet the new worker and discuss the company's safety program. This is also a good time to provide basic safety training as required by OSHA. Personal protective equipment (PPE), such as hard hats, safety glasses, hearing protection, and work gloves, should be provided. New employees should understand that they are expected to use the equipment when working at the job site.

Unfortunately, most small companies do not have a safety director and the responsibility for training new workers falls to the supervisors. Unless they are familiar with training methods, supervisors may need additional training to become effective instructors. Supervisors who are responsible for training workers should work with executive management to obtain information, training materials, and assistance.

XYZ Construction Company
Main Street
Schoharie, NY 12157

Statement of Safety and Health Policy

Employees are our company's greatest asset. We are concerned about our employees' health and well being. Therefore, we must do all that we can to protect our employees from injuries and illnesses.

Management will provide a safe place to work. We will accomplish this by providing safe equipment, proper training, and safe methods and procedures. No job is so important that it cannot be accomplished without injury.

Efficiency depends on the uninterrupted completion of tasks. Accidents interrupt operations. More important accidents cause injury to employees. We must integrate hazard control into every operation. We will comply with every applicable standard and government regulation.

Management and employees must work together for the common goal of preventing accidents and providing a safe place to work. All employees are responsible for safety and health. Safety equipment must not be damaged, removed, or abused. Employees will observe all safety rules and procedures.

Supervisors will see that all rules and procedures are observed by their crews. They are responsible for maintaining safe work conditions.

Everyone must accept an interest in the safety and health program. Together we can control hazards and prevent accidents. We require your full cooperation and help in making our company's safety program successful.

_____
CEO
XYZ Construction Company

Figure 4.1
Employees should be aware of the company's safety policy.

SAFETY PROGRAM INDOCTRINATION

The safety director or other key management person should review the company's safety program and policies with all new employees. To save time and reduce the burden on the individual responsible for the indoctrination, some companies make or purchase videos, or start new employees on a specified day of the week to limit repetition.

During the safety indoctrination, workers should be informed of the following:

- worker's responsibility for safety,
- company safety rules,
- applicable government regulations,
- safe work procedures relative to the employee's job,
- hazard communications program,
- the required personal protective equipment,
- how to report an unsafe condition,
- what to do in the event of an accident,
- what to do in the event of an emergency,
- job site-specific hazards, and
- any other information that increases job-site safety.

JOB-SITE INDOCTRINATION

Job-site supervisors are responsible for ensuring that the worker receives a job-site safety indoctrination. This field indoctrination should include specific information about the job site, including a tour of the job site. During the tour, any hazards associated with the specific job site should be pointed out. For example, the employee may not be aware of some unusual fall hazards or chemical exposures at the job site. In addition, the worker should be told where the first-aid station is, where the material safety data sheets are kept, the type of emergency alarm system used at the job, and where to report in the event of an emergency evacuation.

Informed and knowledgeable employees are safe workers.

Unit 5 – Worker Safety Training and Education

The Occupational Safety and Health Administration requires employers to teach each employee how to recognize and avoid unsafe conditions. Employers must also inform employees of regulations applicable to the type of work the employee will perform. For example, employees that use ladders must be informed of the hazards associated with ladder use, including how to select and set up a ladder, how to secure the ladder, and ladder safety procedures.

MAKING ASSUMPTIONS

To get a job or to impress a new employer, workers frequently overstate their abilities. The mistake a supervisor or foreman can make is to assume that a worker knows how to do the job safely. These assumptions can cause delays, damaged materials, damaged equipment, and, all too often, an injured employee. Never assume that the worker can perform a specific task or operate equipment unless you have verified the individual's competency and knowledge, and are satisfied that the worker can do the job safely. Never assume that a new worker, young or old, experienced or inexperienced, knows how to do the job the safe way.

For example, a company may hire a worker who claims to be a backhoe operator. The worker may have some experience operating a backhoe on a farm, as an independent contractor, or for a contractor who never provided training. Perhaps the worker only operated a backhoe a few times. There is also a good chance that the worker operated a different make and model backhoe than your company uses. The manager must ask the following questions before allowing this worker to proceed:

- Has the worker had any training?
- What level of training has the worker had?

- Did training include safety training?
- What type of experience does the worker have?
- How much experience does the worker have?
- Is the worker familiar with the equipment this company uses?

When satisfied with the answers, the manager can determine whether the worker can operate the equipment to be used. The manager may want to give the worker some specific training or the time to review an operator's manual before proceeding. The backhoe should be in a safe area, away from workers, equipment, and materials, where the operator can demonstrate his or her ability to operate the equipment. Only after the manager is completely satisfied with the worker's abilities, should the worker be allowed to operate the backhoe in the work area.

Workers who operate power tools and small equipment should also be checked-out before being allowed to use equipment. If regulations require a worker to have a license or certification to operate a tool or equipment, the manager must visually check the license or certificate to ensure it is valid and has not expired. For example, the manager should verify that workers who use power-actuated tools have been trained in the use of the specific make and model of tool. If a license, such as a crane operator's license, is required, the manager should make a copy of the license and keep it on file.

SAFETY TRAINING

Safety training is the weakest link in most safety and health programs. A company's safety program should have a safety training plan to indoctrinate and train all managers, foremen, and workers. If the plan is not consistent, some people will be overlooked, and mistakes will lead to accidents that could have been prevented.

If the manager is not satisfied with a worker's knowledge, more training is necessary to ensure that the worker knows how to do the job the safe way, which is the right way. Safety training can be provided in many ways. For instance, a

23

company can provide in-house training at the main facility or it can send workers to schools or seminars. Some companies utilize video training programs.

On-the-job training (OJT) is still the most widely used training method but it tends to be hit or miss, unless the trainer is carefully chosen and the program is well planned. If the manager is not going to train the worker, then someone else must be selected to do the training. The person selected must be a skilled and knowledgeable individual who is familiar with the safety rules and procedures. Before assigning the training task to someone, the manager must be sure the person is willing to provide the training. In addition, this person must be able to communicate the information to the trainee. If used properly, OJT can be an effective way of training new workers.

Computer-interactive training recently has entered the training picture for training new and experienced construction workers. Whatever method is chosen, it must ensure that workers get proper instruction on how to work safely.

Doing the Job the Right Way

Many construction workers have received very little, if any, formal skills or safety training. The construction industry relies heavily on OJT to provide education and training for workers.

Unions have offered apprentice training for many years. This training requires apprentices to complete a prespecified level of classroom training, safety training, skills training, and OJT under the direction of a journeyman for a specific time period. Historically, apprentice training was only available through organized labor.

Open (nonunion) shops account for approximately 75 percent of the current construction work force. Although there are some construction training programs available to non-union workers through vocational schools and local construction associations, availability is limited and many workers never attend any of these programs. The National Center for Construction Education and Research and other organizations are offering training opportunities and providing skills and safety training to workers throughout the country.

Figure 5.1
This worker is not working safely.

If the manager expects workers to follow correct job procedures, they must be trained and have the necessary experience to complete the tasks safely. Every new worker should be evaluated to determine the skill level before being placed on the job. For example, if the employee is a carpenter, the manager should have the worker demonstrate the ability to safely use a skill saw, cut-off saw, radial arm saw, or other equipment that the employee might be expected to use at one of the job sites. Some of the things the manager should verify include that the worker knows how to lift safely, uses safety glasses, inspects the equipment before use, uses machine guards, and maintains a clean work area. A checklist of tasks the carpenters should be able to perform would be helpful.

Key to Preventing Accidents

The key to preventing accidents is **knowledge** of the safe way to do the job. Safety knowledge is acquired through a combination of education, training, and experience. Supervisors and foremen must be familiar with the work they supervise. To prevent accidents, they must also be aware of applicable safety regulations and any specific safety precautions that must be taken.

For example, foremen responsible for workers that use scaffolds should be aware of OSHA's scaffold regulations, the manufacturer's set-up procedures, and any job-specific hazards that may exist. Foremen should take every opportunity to learn about safety and applicable safety regulations.

Because most accidents are caused by the unsafe acts of workers, supervisors and foremen should share their knowledge with workers. Workers must know the applicable regulations and how to do what needs to be done safely. For example, all workers must wear safety glasses when operating power saws; and, the guards on machines should not be blocked or removed. The proper tools and equipment must be available and used to complete the task safely.

Workers should not be allowed to use unsafe tools or equipment or to perform unsafe acts, even for a moment. Many workers have been seriously injured or killed when a task that was only going to take a moment led to an accident. Workers have been buried in unprotected trenches when trying to retrieve a tool left in the trench. Workers have died from falls off the top of ladders while doing something that was only going to take a few seconds, like changing a light bulb. Workers have been electrocuted while using a defective electric drill just to drill one more hole. Nobody should be allowed to take chances. It is like playing Russian Roulette.

Toolbox Safety Talks

The most common method of sharing safety information in the construction industry is the toolbox safety talk, sometimes referred to as a tailgate safety talk. Toolbox safety talks are generally given weekly. They should be used to remind workers of the hazards they may encounter and how they can protect themselves from the hazards. The talks are generally brief and to the point and should not be considered a substitute for proper safety training.

Toolbox safety talks require some preplanning by the foreman or supervisor who will lead the discussion. The leader should take it seriously and make sure the crew knows

Monthly Tool Box Talk

Trench Safety

Working in a trench is one of the most hazardous jobs in construction. Hundreds of people die and thousands are seriously injured each year due to cave-ins.

Soil weighs between 90 and 140 pounds per cubic foot. Therefore, one cubic yard (3 ft x 3 ft x 3 ft) can weigh as much as a pick-up truck. If a person is buried, there is not much chance of survival.

There are many things that affect soil stability, such as the type of soil, water, and vibration. Soils saturated with water and previously disturbed soils are very dangerous to work in. But, don't be fooled, even hard soil and rock that appears stable can cave in.

Before entering a trench, the competent person at the job site must inspect the trench and the protective system to ensure that the trench is safe to enter. Because, there are recorded incidents of people buried and killed in trenches three to four foot deep, even shallow trenches must be inspected by a competent person before entering.

Let's review some basic trench safety tips.

- Before digging, locate underground utilities.
- Enter all trenches only when sloped, shored, or shielded.
- Never go outside the protective system, not even for a moment.
- Eliminate or control water accumulation before entering.
- Be extra alert when working in or near previously disturbed soil.
- Do not permit vehicles too near to the edge of the trench.
- Check regularly for hazardous materials and oxygen levels in the trench.
- Spoils, tools, materials, and equipment must be stored at least two feet from the edge of the trench.
- Never allow machines to run unattended.
- Use a ladder or ramp to get in and out of the trench. Never ride in the backhoe bucket or on a crane hook.
- Never climb on shoring or shields.
- Hard hats must be worn when working in and around trenches.
- Stay out from under suspended loads.

Approximately 50 percent of the people killed each year in trenches die trying to rescue someone else who has been buried by a cave-in. If somebody is buried, do not jump in the trench to try to save him or her until you are sure the sides of the trench are supported and safe. You should not attempt a rescue, unless you have been properly trained in trench rescue techniques.

Remember, if you are buried by a cave-in, your chance of survival is very low. Therefore, always be sure the trench walls are sloped, supported, or shielded and that the trench is safe before you enter.

Figure 5.2
Sample tool box safety talk

that it is serious. A simple outline should state what is to be covered. Goals, objectives, a list of important information to cover, and a brief recap should be included.

In other words, the leader should be prepared to do the following when presenting the information to the workers:

- tell them what will be discussed,
- present the information and encourage discussion, and
- summarize what was discussed.

Adults do not like to be preached to. The leader should show the worker how the information being presented is relevant to their daily work routine. Examples of situations should be given. The workers should be involved, and they should show how they have applied their knowledge and experience to the problem. To keep their attention, the leader must not insult their intelligence or read from a script.

If the leader is not sure of what to discuss or runs out of ideas, the company's safety director or management might be able to help. Some companies provide prepared safety talks or videos for safety toolbox talks. This makes the job a little easier, but the information must relate to the job site. Interaction and discussion should be encouraged. The information should be presented in a way that all workers understand it. The leader should be prepared to answer questions.

Before letting workers return to work, the leader should get some feedback from the group to ensure that the message was understood. For example, the workers could describe any relevant hazards at the job site and make suggestions to correct or eliminate relevant hazards. The workers can demonstrate the knowledge they acquired. Can they demonstrate the skills they were taught? Do they have the right tools and equipment to do the job safely? Do they have the right attitude?

JOB HAZARD ANALYSIS

Job hazard analysis (JHA) is used to identify and evaluate job hazards and training needs. The analysis is an effective, organized method of studying a job to develop the safest, most effective ways to accomplish a task.

Which jobs to analyze should be determined by considering which jobs have the greatest potential for serious injury, have a history of lost workdays due to accidents, are most frequently performed, are performed by the greatest number of workers, and are new jobs. All tasks are evaluated and procedures are developed for performing different tasks. For example, step-by-step procedures are developed for setting up a ladder, cutting 2×4s with a cut-off saw, hoisting materials, installing guard rails, and so on.

By creating procedures for a specific task (e.g., hoisting materials), the hazards that may occur during performance can be identified and evaluated. After all potential hazards are identified, methods to control or eliminate each hazard can be established. This information can be used to improve job-site conditions, select tools and safety equipment, and train employees.

Benefits of using JHA are

- identification and control of hazards that cause accidents,
- consistent job and safety training,
- improved worker skills and safety performance,
- increased management awareness of hazards,
- increased employee involvement,
- more accurate accident investigations, and
- compliance with regulations.

There are two basic ways to perform a JHA: direct observation and group discussion. Some companies combine the different methods.

The **direct observation** method requires the cooperation of a worker while the observer records the job steps as they are performed. Potential hazards also are recorded. Video cameras can be used to record the task for more detailed review later.

The **group discussion** method involves several people who are familiar with the task. The group meets and members draw on their experience to help analyze the job.

No matter which method is chosen, the following steps must be performed:

(1) List job steps in sequence. Record enough information to describe each action that must be taken. Be brief and use active verbs (e.g., lift, push, move, carry). When possible, break the job into fewer than ten steps.

(2) List all potential hazards connected with each step. For example, determine if there is a possibility of one or more of these following exposures: eye hazard, fall hazard, overexertion, unguarded equipment, chemicals, being struck by or caught in something, and poor ventilation. Ask what could cause an injury or illness to workers or the public, and what could damage materials, equipment, or the environment?

(3) List ways to eliminate or control the hazards associated with each step. The hazards must be reviewed carefully to find ways to eliminate or control them. Hazards can be minimized by:

- engineering controls;
- administrative controls;
- providing personal protective equipment (e.g., safety glasses, hard hats, gloves, respirators);
- changing the sequence of steps;
- providing machine guards;
- improving ventilation;
- using different tools or equipment; and
- other controls and methods, depending on the job.

The completed JHA should always be reviewed by managers and workers to ensure that nothing has been omitted. After final approval, the document can be used to establish a standard procedure and to train workers. JHAs are often incorporated into a company's standard operating procedures and safety program.

Sequence of Basic Job Steps	Potential Accidents or Hazards	Recommended Safe Job Procedure
Break the job down into its basic steps, e.g. what is done first, what is done next, and so on.	For each job step, ask yourself or the group what accident could happen to the person doing the job or someone else in the area where the work is done.	For each potential accident or hazard, ask yourself how should the workers do the job step to avoid the potential accident, or what should be done or not done to prevent the accident?
Do this by: (1) observing the job (2) discussing it with the operator (3) drawing on your knowledge of the job, or (4) a combination of the three.	Get answers by: (1) observing the job (2) discussing the job with the worker (3) recalling past accidents, or (4) a combination of the three.	This can be done by: (1) observing the job for leads, (2) discussing precautions with the group and/or experienced workers, (3) drawing on your experience and the experience of others, or (4) a combination of the three.
Record the job steps in their normal order of occurrence. Describe what is done, not the details of how it is done. Usually three or four words are sufficient to describe each basic job step.	Ask yourself or the group: Could someone be struck by or contacted by anything; could someone strike against or come in contact with anything; could someone be caught in, under, or between anything; could someone fall; could worker(s) overexert themselves; is anyone exposed to anything injurious, such as gas, vapors, fumes, welding rays, and so on?	Be sure to describe specifically the precautions a worker must take. Do not leave out important details. Number each precaution with the same number you gave the potential accident or hazard that the procedure seeks to avoid.

Figure 5.3. Sample job hazard analysis form (Source: *Today's Supervisor*, National Safety Council, Itasca, IL)

31

Job Hazard Analysis

Job:	Ladder Set-up		
Date: June 9, 199X	Analysis by: Joe Smith		Approved by: John Doe

Job Steps or Sequence	Potential Hazards	Recommended Safe Procedure
(1) Select and inspect ladder	(1) Cracked, damaged, rails, feet, rope, oil, grease. Wrong size.	(1) Damaged ladders should be tagged and removed from service. Check list.
(2) Move ladder to work area	(2) Lifting, tripping, striking objects or people.	(2) Two people required. Use clear path.
(3) Inspect set-up area	(3) Well lighted, soft surface, wires overhead, displacement, clear area. Weather conditions.	(3) Remove debris, set on level, firm ground. Set up away from aisles, roads, doorways, and so on.
(4) Stand ladder up	(4) Lifting, flip-over, wires.	(4) Use two people to lift and stand ladder. Do not set near wires or access paths. Set ladder up inside during bad weather.
(5) Extend ladder	(5) Strains, rope burn, improper angle.	(5) Wear gloves rest on rigid support extend 3 ft above point of access. Angle should be approx 4V:1H.
(6) Check stability	(6) Ladder could tip over.	(6) Attempt to move ladder with hands. Climb on first or second step.
(7) Climb ladder	(7) Falling or slipping.	(7) Hold ladder for first person up. Maintain three-point contact.
(8) Secure ladder	(8) Tip over, debris at point of access.	(8) Secure ladder to structure by tying off or blocking. Clean the access area at top.
(9) Climb ladder	(9) Falls while climbing and securing ladder.	(9) Hold extended ladder rail when climbing back onto ladder. Use three-point contact when climbing back down.

<u>Figure 5.4.</u> Sample JHA for ladder setup

ADDITIONAL TRAINING TIPS

For training to succeed, managers at all levels must have positive attitudes. The trainers, including foremen and supervisors, must understand the training purpose and importance. Trainers must also have a positive attitude and be enthusiastic.

Establish a regular schedule for training seminars and programs. When training is scheduled, there should be no excuses for not attending. The best results occur when everyone attends and participates in the activity. Full participation ensures that everyone gets the same message.

When possible, the manager should provide training in the workers' native language so they clearly understand the message. The level of training should be geared to the audience, and the message should be presented in an interesting manner. Variety can be added by using different types of visual aids. For example, the use of video tapes for one message, slides at the next session, and hands-on applications for skills learning is helpful. A quiz or some form of feedback exercise at the end of the session should be given. The workers should demonstrate the skills they have learned. Whenever possible, some form of handout material for the worker to take home should be provided.

The manager should not give "lip service" to training. Training sessions should not be held just for documentation. Without training, the best safety efforts can be reduced to a waste of time, money, and energy.

Unit 6 – Safety Rules and Procedures

Every company must have rules, and some of the most important rules are safety rules. Procedures and rules provide direction, so workers know what they should and should not do. Rules are used to establish proper performance, and contrary to the cliché, rules are not made to be broken. What would the roads be like on the way to work everyday if there were no driving rules or if everyone broke the rules of the road?

BASIC SAFETY RULES AND PROCEDURES

Basic safety rules and procedures should be established so all employees know what is expected of them. Rules are generally a list of do's and don'ts that direct acceptable conduct. Procedures are established to provide specific directions or methods for accomplishing a task. The supervisor or foreman must know the safety rules and procedures to educate and train workers.

Job-Specific Safety Rules

Similar to basic rules, job-specific rules help workers understand what they should and should not do. The difference is that job-specific rules are unique to a particular job site or task. For example, crane operators have job-specific safety rules that apply only to crane operations. The rules that apply to laborers laying pipe in a trench are different than the rules that apply to laborers handling materials during building construction.

Compliance

Management and labor must comply with all company rules. Local, state, and federal laws require compliance with some rules. For example, compliance with applicable OSHA regulations is mandatory. State plan states have adopted the

XYZ Construction Company
Main Street
Schoharie, NY 12157

General Safety and Health Rules

(1) Good housekeeping shall be maintained at all times.

(2) Smoking is not permitted in work areas where combustibles, flammables, or chemicals are stored or handled.

(3) Hard hats and safety glasses shall be worn at all times on the job site. Other personal protective equipment, such as respirators, gloves, hearing protection, harnesses, and lifelines shall be worn when applicable.

(4) Loose clothing and jewelry shall not be worn when operating machinery and equipment.

(5) The use and possession of drugs and alcoholic beverages on the job site is forbidden. Any employee suspected of being under the influence of drugs or alcohol will be required to report immediately for a substance abuse screening.

(6) Do not engage in horseplay; avoid distracting others; be courteous to other workers at all times. Violent actions will not be tolerated.

(7) Never enter into restricted areas unless authorized.

(8) Do not take chances. If you are not sure how to perform a task safely, ask your supervisors for assistance and instructions.

(9) Immediately report all unsafe conditions or situations to your supervisor.

(10) All accidents or injuries, no matter how minor, must be reported to your supervisor immediately.

<u>Figure 6.1</u>
Basic safety rules

OSHA regulations or created regulations that are at least as demanding.

Managers and foremen should set a good example by complying with the rules. No exceptions should be made. Occasionally, the manager may need to take action to ensure that all safety rules and work procedures are followed.

For example, a visitor or even a manager may enter the work area without a hard hat. The supervisor should remind the person that a hard hat is required and that he or she may not enter into the area without one. People who honestly care about safety will be pleased that the supervisor is concerned about their safety and took the initiative to remind them of the hazard.

Enforcement

The foreman or supervisor must ensure that workers are aware of job-site rules and must enforce them. Like a police officer, the manager must enforce the rules to ensure the safety of the crew. If violations are overlooked and workers are permitted to disregard the rules, accidents will happen.

DISCIPLINARY ACTION

Managers, foremen, and supervisors are responsible for taking disciplinary action when workers violate the rules. Workers who do not follow rules that have been established to protect them and their coworkers endanger themselves and others. If a violation is observed, one of the following actions must be taken immediately.

- **First Warning:** The first time an employee violates a rule, he or she should be reminded of the rule and given a verbal warning. It is a good practice to document that the employee was informed of the violation. Make note of the day, time, location, violation, and corrective action.

- **Second Warning:** The second warning should be a verbal warning accompanied by written warning. A copy of the written warning should be given to the employee and to the manager or safety coordinator. A

Notice of Disciplinary Action

Employee Name:
Job Title:
Job Site and Location:
Date and Time of Occurrence:

Check one of the following.

First offense___ Second offense___ Third offense___ Fourth offense___

This is to advise you that you are hereby placed on notice that you were observed violating the following company rule, policy, or procedure.

Description of the rule, policy, or procedure that was violated.

_____

_____

_____

_____

_____

Describe the corrective action that was taken to prevent the violation previously described from occurring again.

_____

_____

_____

_____

_____

First offense: Verbal warning and supervisor's record maintained. Second, third, and fourth offenses: (1) Verbal warning. (2) Written warnings: copy provided to employee; copy placed in employee's personnel file; copy provided to union representative where applicable. (3) Follow-up action in accordance with company disciplinary action policy.

Supervisor's signature:_____ Date:_____

Figure 6.2
Sample disciplinary action form

copy of this warning must be placed in the employee's personnel file. Some companies require the employee to meet with the safety coordinator for counseling.

- **Third Warning:** The third warning is similar to but more serious than the second warning. The worker must be advised of the violation, and the warning must be documented. A copy of the written warning should be given to the worker, safety coordinator, and top management. A copy should be placed in the employee's file. A meeting with the employee, foreman, safety coordinator, and top management should be held to determine why the employee is not willing to comply with the company's rules. Top management must determine what action will be taken at this time.

- **Fourth Warning:** Employees that accumulate four warnings in a twelve-month period (or other pre-established time frame) should receive a written notice of the violation and be required to meet with top management to determine the action to be taken. Employees who are not willing to follow the rules, especially after being warned several times, are a threat to themselves and their coworkers. Therefore, termination of employment may be an appropriate action at this time.

Unit 7 – Substance Abuse

Substance abuse is rampant throughout the United States. Unless a company has a substance abuse program, drug and alcohol abusers probably work there.

Statistics show that as much as 20 percent of the work force will test positive for illegal drugs. In addition, 80 percent of all illegal drugs sold are sold in the work place. Anyone at a job site could be a potential drug dealer.

Any worker of any race, socioeconomic background, or occupation could be a substance abuser. It is not always the substance abuser who pays the price for abuse. The use of legal or illegal drugs, including alcohol, in the work place can place workers in danger. Substance abuse can affect job performance even if it occurs outside of the work place.

Drugs can:

- cause workers to take more chances;
- increase the potential for injury to the abuser or a coworker;
- increase absenteeism and lateness;
- decrease productivity and quality of workmanship;
- increase a company's health insurance and workers' compensation costs;
- increase theft of materials, tools, and equipment; and
- increase a company's financial losses and liabilities.

The most commonly used drugs are alcohol, marijuana, cocaine, PCP, codeine, demerol, and valium.

TYPES OF DRUGS

Alcohol

Alcohol, although frequently not referred to as a drug, is the nation's most commonly used drug. It can be purchased legally in most parts of the country. When used in the work

39

place, alcohol can be very dangerous. Alcohol use causes impaired judgment, loss of concentration, and slower reflexes. It is also a depressant.

Alcoholic beverages should never be permitted in the work place. Although a couple of beers at lunch may seem harmless, they could lead to serious accidents or even fatalities.

Hangovers are often the cause of absenteeism and lateness, which increases the burden on other crew members. Following a late night out or a "liquid lunch," workers often arrive at work impaired. If a worker's breath smells of alcohol or if a worker is suspected of being impaired upon arrival at work, the supervisor should handle the situation in accordance with the company's substance abuse policy and procedure.

Illegal Drugs

Marijuana, cocaine, PCP, and other illegal drugs are sometimes more difficult to detect than alcohol, so foremen and supervisors must become familiar with the signs and symptoms of drug abuse.

Marijuana slows a person's physical reflexes, impairs short-term memory, and reduces mental powers. It also affects a person's judgment of time, space, and distance.

Cocaine is addictive. This substance increases the user's sense of well-being and causes overconfidence.

Stimulants, often referred to as speed or uppers, cause a person to work quickly and carelessly. Other dangers include hallucination, confusion, depression, paranoia, and addiction.

Depressants, often referred to as downers, include tranquilizers, sedatives, and barbiturates. When used without a prescription, valium is illegal.

Narcotics, such as heroin, codeine, or demerol, are addictive and lead to complete disregard for safety or anything else except more drugs. The high cost of these drugs often leads to theft and crime to support the habit.

Hallucinogens, referred to as designer drugs, LSD (acid), PCP, psilocybin, mushrooms, or peyote, dangerously affect a person's mental functions. This often leads to unpredictable behavior, emotional instability, and sometimes violent behavior.

Inhalants, such as nitrous oxide, aerosols, and airplane glue, affect the central nervous system. This could result in dizziness or lack of coordination.

Prescription and Over-the-Counter Drugs

Many prescription and over-the-counter drugs contain controlled amounts of some of the drugs previously listed. Even when taken as prescribed, they can cause ill effects. When inappropriate quantities are taken, the effects are intensified.

Drugs of this type have warnings on the packaging that advise against driving or operating equipment. Employees should be reminded to take these warnings seriously.

SIGNS AND SYMPTOMS OF ABUSE

The signs and symptoms of alcohol use are well known. Workers who have been drinking may smell of liquor or beer. If they have ingested a large quantity, they may stagger, talk too much, exhibit slurred speech, or experience loss of some inhibitions.

Marijuana, when smoked, smells like burning leaves, which can be detected on the person's breath, clothing, or in the air where it was used. The effects may cause reddening of the eyes, often accompanied by watery, light-sensitive eyes. The person's coordination may be impaired and short-term memory also may be effected. The person may also experience spurts of laughing and giddiness.

Cocaine users may talk a lot or exhibit a sense of exhilaration, euphoria, and overconfidence. Regular users may show signs of redness under the nostrils and may tend to duck out of sight every 20 to 30 minutes for a quick "fix." They also may exhibit signs of depression, nervousness, or irritability.

Tranquilizer users may exhibit some sense of well-being and may lose some inhibitions or show signs of being drunk. Higher doses may cause the worker to be in another world, mentally confused, or physically unstable. They would appear to have a loss of memory, or be drowsy, uncoordinated, or disoriented. In some cases, the individual may become enraged or show signs of altered personality.

Signs of depressants use are similar of those of alcohol use. The person will appear to be relaxed, sedated, or drowsy. The person may be impulsive, jovial, and sociable. Higher doses lead to inability to react quickly or perform precise skills. The individual may also alternate between euphoria and hostility or aggressiveness.

Stimulants widen the pupils of the eyes, increase respiration, depress appetite, and decrease fatigue. Even moderate doses can cause adverse effects, such as agitation, lack of concentration, anxiety, confusion, blurred vision, tremors, and mild depression.

Inhalants produce effects similar to those of alcohol use. Higher doses produce laughing and giddiness, dizziness, illusions, and, in some cases, psychedelic effects.

Heroin users tend to avoid labor-intensive jobs; however, they occasionally find their way onto construction sites. Intravenous use or skin popping leaves "track marks" or lesions in the skin. Other signs of use include reduced breathing, constriction of the pupils, increased perspiration, nausea, and sometimes vomiting. Withdrawal symptoms may resemble symptoms of flu, such as uneasiness, tears, and a runny nose.

HANDLING SUBSTANCE ABUSE PROBLEMS

All supervisors play a very important role in controlling substance abuse at the work place. If the supervisor detects an employee who appears to have a substance abuse problem, he or she should check the company's substance abuse policy and consult with the personnel department, safety director, or appropriate individual before taking any specific action. Confidentiality is of utmost importance. Suspicions should be discussed only with individuals who need to know.

If the suspected worker is in a position where harm could be caused to self or others, the manager should make up some excuse to temporarily relieve the worker from assigned duties until the manager can consult with the appropriate persons. There are many legal issues involved, so the worker should not be not accused or reprimanded. Managers should not attempt to be substance abuse counselors; it is not their job. If the worker needs help, the company can arrange to send the employee to an employee assistance program (EAP) for professional help.

Unit 8 – What to Do If an Accident Occurs

Emergencies take many different forms. What should the supervisor do if someone is injured? If a fire starts? If a worker falls from an elevated surface and is hanging from a lifeline? If a worker is buried in a trench? If a worker is unconscious in a confined space? What if there is a chemical spill at the job site? These are just some of the emergencies for which the supervisor must be prepared. Everyone hopes an accident will never happen. But if it does, would a crew know what to do?

EMERGENCY ACTION PLANS

If a job site does not have an emergency action plan, panic will follow; emergency action will be significantly delayed. Federal and state safety standards require contractors to have emergency action plans for all job sites. The plans must be based on the types of emergencies that could occur at the job site. For example, underground contractors must be prepared to implement a rescue plan in the event of a cave-in, and bridge contractors must be prepared to rescue workers after a fall into a safety net.

OSHA requires contractors to have a written job site-specific emergency action plan. The parts of the plan that the employee must know must be reviewed with each employee upon initial assignment to the job site. All employees know what to do in the event of an emergency. Employers with fewer than eleven employees are not required by OSHA to have a written plan, but emergency procedures must be communicated to all workers. A written plan will help ensure that procedures have been established and that everyone will know what to do.

Some state and local laws have different requirements. The supervisor must determine what is required and implement emergency plans that comply with these standards. If the requirements are not clear, workers should check with their employer or company safety coordinator.

Emergency Procedures

Emergency procedures must prescribe action that must be taken by managers, foremen, supervisors, and employees in the event of an emergency. At the very least, the procedures must include the following:

- Who and how to call for help. For example, notify the foreman or call the fire department, use the telephone, call the posted telephone number.
- Method of sounding the alarm. For example, what is the signal (alarm bell, whistle, air horn, and so on) to evacuate the work area in the event of a fire or other emergency?
- Escape routes and procedures in the event of an evacuation. For example, evacuate the area by using the stairway on the south end of the building.
- Method for accounting for all workers after an evacuation. For example, where to meet after they get out, who to report to, who to report to if someone is not accounted for.
- Who is the designated first-aid provider? For example, the foreman or a designated individual may be responsible for providing first aid.
- Location of the first-aid kit. For example, first-aid supplies may be kept at the first-aid station, in the job site trailer, or in the gang box.
- Method for rescuing workers in the event of a specific type of accident. For example, how will an unconscious worker be evacuated from a manhole, or how will a worker suspended by personal fall arrest equipment be rescued or lowered to the ground?
- Who will perform the rescue? For example, an on-site rescue team trained in confined-space rescue or an off-site rescue squad trained in high-level rescue. (**Note:**

<u>Figure 8.1</u>
Do not waste precous time. Call for help if an accident happens.

Prejob planning must include making sure the specified rescue team on- or off-site can handle the type of rescue that could occur and that the necessary equipment is available. Is their response time appropriate for the type of rescue?)

- What to do if a chemical spill occurs. For example, report the spill to the project manager and notify the fire department or chemical spill clean-up team.

- Any other procedures necessary to control any other emergency situations that could reasonably be expected to occur at the job site.

- Above all, have a plan, do not try to be a hero, and do not panic. Remember that accident prevention is the first line of defense.

MEDICAL SERVICES AND FIRST AID

If an employee is injured, prompt medical attention must be available. Provisions for medical attention should be established before starting the job. In fact, job-specific procedures for first aid and medical services should be included in the job site emergency action plan.

In the absence of an infirmary, clinic, hospital, or physician to treat injured workers, a person certified in first-aid training must be available on-site. A court decision has determined that the emergency response time must be four minutes or less. Therefore, most job sites are required to have a person trained in first aid on-site, unless the local paramedics or rescue team can arrive at the job site in less than four minutes' travel time.

An appropriate first-aid kit must be available in a weatherproof container. First-aid kits must be inspected at least weekly to ensure that expended items are replaced. The contents of the items in the first-aid kit must be kept in individually sealed packages.

Job-site emergency plans also must include a written exposure control plan for bloodborne pathogens, which are microorganisms in human blood, such as hepatitis B virus and human immunodeficiency virus (HIV). People designated to provide first aid must receive training in how to protect themselves from and control exposure to bloodborne pathogens. Specific methods must be established for preventing human exposure and for the removal, handling, decontamination, and disposal of contaminated personal protective equipment and first-aid supplies following an exposure incident. Any areas or equipment exposed to blood also must be decontaminated according to the procedures established in the written exposure control program. Details are available by referring to a company's written exposure control program.

Equipment to transport victims or access to an ambulance service must be available. However, injured victims should not be moved, unless the victim's well-being is threatened by the immediate location. Victims should only be moved by people trained and qualified in the packaging and transport of injured victims. An example of such a trained person would be an emergency medical technician. Moving a victim who is not properly prepared for transport can increase injury severity.

Where workers' eyes or bodies could be exposed to harmful corrosive materials, a suitable eyewash and emergency shower for quickly drenching or flushing of the eyes or body with fresh, clean, tepid water must be available in the work area. Quick, immediate response to this type of exposure can prevent serious chemical burns and eye injuries.

Unit 9 – Accident Investigation

What is an accident? An accident is an unplanned and unwanted event that may have resulted in an injury, illness, property damage, loss of productivity, or damage to materials. Accidents do not just happen, they are caused.

WHY ACCIDENT INVESTIGATIONS ARE PERFORMED

An accident investigation is a system for determining the facts that lead up to and caused the occurrence. With this information, future accidents of a similar type can be prevented. All accidents should be investigated to:

- identify the cause of the accident,
- determine the facts,
- meet recordkeeping requirements,
- determine if there were any changes or deviations from normal procedures,
- create a database for analysis, and
- have information to help prevent future accidents.

The investigation should be used to determine the facts about what happened and not to fix blame. Although foremen and workers should be held accountable for their actions, it is more important to correct any mistakes so they are not repeated.

ACCIDENTS HAVE CAUSES

Accidents are caused by one of two categories of causes: unsafe acts and unsafe conditions.

Unsafe acts are violations of commonly accepted safe practices and procedures, such as:

- something a person did that he or she should not have done (i.e., unauthorized use of equipment);

49

- something a person should have done differently (i.e., failure to follow established procedure); or
- something a person failed to do that should have been done (i.e., properly securing a load).

Unsafe conditions are the physical or environmental conditions that directly caused or permitted the occurrence of the accident. These are often the most common, such as:

- defective or damaged tools or equipment,
- poor housekeeping,
- no personal protective equipment available,
- inadequate or unguarded machines or equipment,
- improper storage of materials, and
- uncontrolled contaminants in the environment.

Accident investigations often reveal that a combination of unsafe acts and unsafe conditions caused the occurrence. For example, an employee may knowingly use a defective tool. The unsafe condition is the defective tool; the unsafe act is the intentional use of that tool. If one or the other cause was eliminated, the accident would not have occurred.

Unsafe acts and conditions are the obvious causes of accidents. However, when performing the investigation, it is very important to determine the root cause of the accident, which may be tracked back to actions performed or not performed by other employees. Management activities or the lack of management activities also contribute to accidents. Examples follow:

- lack of a safety programs, rules, and procedures;
- lack of employee safety training;
- lack of supervisor safety training;
- inadequate job planning, poor job-site layout, or lack of tools or equipment;
- failure to ensure proper maintenance of tools and equipment;
- failure to enforce safety rules and regulations; and
- failure to establish and enforce safe procedures.

PERFORMING THE INVESTIGATION

When an accident occurs, the first thing that must be done is to rescue the victim and to obtain medical treatment. After the victim has been attended to, the scene of the accident should be secured until the accident investigation is completed. Do not disturb the accident scene, and keep unauthorized spectators out of the area. To secure the accident scene, the area should be roped off, taped off, locked up, or designated as a restricted area. If the scene of the accident is not secured, conditions will change quickly and a thorough investigation cannot be completed.

Before anything changes at the scene of the accident, pictures should be taken. Keeping a disposable camera or inexpensive instant picture camera at every job site is a good practice. Diagrams of the scene also should be made. Thorough observation of the area should identify what was in the area. Also, list the names of potential witnesses, who was in the area, and who may have observed what took place just prior to the accident. Ask these individuals not to discuss the accident with anyone until you have a chance to discuss it with them.

Depending on the injured person's medical condition, the injured worker should be interviewed as soon as possible. Potential witnesses should also be interviewed promptly before their account of what happened changes. It is important to remember that a person did not actually have to see the accident to be of value to the investigation. Knowing what took place and what was in the work area prior to the accident is also important to the investigation.

Interviewing the Injured and Witnesses

Each potential witness should be interviewed separately without being disturbed. If an office or job-site trailer is not available, the meeting could be held in a car or pickup truck. Above all, the interviewer should not try to obtain a group account of what occurred. Following are interviewing points:

- Each individual should be met with separately and promptly.

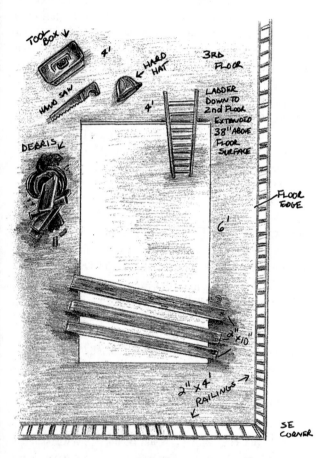

Figure 9.1

Know what the accident scene looks like. Make a diagram.

52

- The person should be reminded of the purpose of the investigation and informed that the reason for the interview is to find the cause, not to fix blame.

- Each individual should be asked to describe what was observed prior to and when the accident occurred. The witness should not be interrupted while providing an account of what was observed. The interviewer should not write while the witness is talking; if the statement is being recorded, the witnes should be asked for permission before starting.

- When the witness is done talking, questions may be asked to fill in the blanks.

- The interviewer should review his or her understanding of what was observed with the witness to ensure an understanding of what the person just stated.

- The interviewer should ask the witness to discuss his or her thoughts on what may have been done to prevent the accident. These ideas may be valuable when it is time to take action to prevent similar occurrences in the future.

During the interview, the witness should never feel threatened. The interviewers should never be sarcastic or make accusations.

After the interview is finished, the injured person and the witnesses should complete the accident investigation form. Answer all questions based on the facts that you have obtained. Try to complete all parts of the accident investigation form. Sections that can not be answered based on fact should be left blank or marked with a question mark until factual answers can be obtained. All information that is written down can be used in a court of law or hearing, which is why all the information that is documented should be factual.

A company's accident investigation procedures should be checked to determine exactly to whom the investigation forms and reports should be sent. They should be treated as confidential documents and should not be distributed or copied.

Accident Investigation Form

EMPLOYER:		JOB SITE:	
Employee:		Age:	Sex: M or F
Occupation:		Length of Service:	
Incident Date:	Time: AM / PM	Department:	
Who advised you of the incident?		When:	

Description of incident:

Nature of injury and part of body involved:

What protective equipment was being used?

What job was employee doing? Include tools, machine, and materials used.

What events occurred immediately before the incident?

What did the employee do unsafely?

What safety rules were violated?

What was defective, in an unsafe condition, or wrong with the method?	
What safeguards should have been used?	
What steps have been taken to prevent similar incidents?	
What steps should be taken to prevent a recurrence?	
Was any property, product, or equipment damaged? If yes, describe:	
Who provided medical care?	When:
Doctor:	Hospital:
Did employee report to work the next scheduled work day? YES NO	
Names of witnesses to incident:	

SIGNED: _____ Supervisor	DATE: _____

Figure 9.2
Accident investigation forms can help you collect the facts.

CORRECTIVE ACTION

Unfortunately, the information obtained from an accident investigation is after the fact. However, this information can be used to prevent a similar occurrence. Therefore, corrective action should be taken immediately when possible.

Unit 10 – How to Handle the Media

An accident happens; the manager calls for help; the alarm is sounded; help is on its way and so is the media. Local television stations and newspaper reporters are tuned in to fire and police radio frequencies. When something of interest grabs their attention, they respond to the scene. The supervisor must know what to do when the media starts asking questions.

WHO SHOULD HANDLE THE MEDIA?

A company policy for handling the media should be established. Foreman and supervisors should be familiar with this policy and should be instructed what to do if the media approaches them.

Companies should assign somebody to handle the media if an accident occurs. The company's spokesperson should be educated in the appropriate methods for handling the media tactfully. Under normal circumstances, this individual would not be the job-site foreman. However, a foreman or supervisor at the job site may be approached by reporters, so it is important to know some basic rules for handling the media.

Never Say "No Comment"

Reporters are determined to get a story. When the spokesperson is approached by reporters, the best thing to do is give them what they want, which is answers. Otherwise, they will find other sources of information, which might not be accurate.

Reporters or photographers should not be allowed to roam around the job site unsupervised. An area should be established for the media to wait where they will not be injured or get in the way. All briefings are to be held at this specified location.

The spokesperson must never respond to a reporter's questions by saying "no comment." The spokesperson may appear

57

to be hiding something and encourage the reporter to search for answers from somebody else. Telling a photo journalist not to take pictures is the same as saying "no comment."

The foreman or supervisor must be prepared to be the temporary spokesperson until the company's spokesperson arrives at the job site. The statement could read as follows:

My name is Joe Smith, and I am the foreman for XYZ Construction Company. Because of the rush of the emergency, I do not have any verifiable information at this time. Please give me 45 minutes to gather some facts for you. In the meantime, please stay in this safety area until I return.

A statement of this type "buys some time" for the spokesperson to arrive and gather the facts. It also shows the media the information they are looking for will be released as soon as it is available. If the spokesperson does not arrive, the foreman or supervisor may have to talk to the media again, in 45 minutes. It may be possible to buy some more time by telling them information is still being collected.

If the company does not send a spokesperson, someone should talk to the media. If possible, basic information about what happened should be provided. No comments should be made about who was at fault or speculate how the accident happened. The foreman or supervisor should tell them he or she does not know something if that is the case.

Following is an appropriate response to an accident involving a man who fell from a roof:

My name is Joe Smith and I am the foreman for XYZ Construction Company. A crew of workers was installing roofing material, and a worker fell from the roof. He survived the fall, but the extent of his injuries are unknown at this time. The emergency medical technicians are with him and will transport him to the hospital for treatment as soon as possible. At this moment I do not know any other details, but I will return when more information is available.

The statement should be kept brief. The media is looking for a story in 25 seconds or less. Small tape recorders and camcorders can be recording what is said at any time, even if they appear to be turned off. Everything said could be recorded. Nothing is off the record.

When possible, clear any statements with management and legal council before making the statement to the media. For protection, a written copy of the statement to the media should be provided.

If the media is told somebody will return at a specific time to brief them, this appointment must be kept even if additional facts are not available. It is important the media believe they are not being avoided or information is not being withheld. They might try to obtain additional information from workers and other people who may have been on or near the job site when the accident occurred. Information from other sources is often speculative and speculative information can damage a company's reputation.

Points about Handling the Media

Following is a list of points the spokesperson for a company should pay heed to when responding to the media:

- Always tell the truth.
- If the facts are not known, no speculations or predictions should be made.
- Nothing should be said off the record.
- The spokesperson should never say "no comment."
- The statement should be brief and to the point.
- The media should be restricted from wandering around the job site.
- Sunglasses must be removed when speaking to the media.
- The spokesperson should remove gum, candy, tobacco, and so on from the mouth before speaking.
- Response should be quick. The media must not be ignored or avoided.

- Accusations or speculations about who may be responsible should not be made.
- Only accurate information should be given.
- Time must be allowed for the company spokesperson to arrive at the job site and to collect the facts.

Unit 11 – How to Handle an OSHA Inspection

CHECKING CREDENTIALS

An OSHA inspection begins when the compliance officer arrives at the job site. In fact, some compliance officers will stand in a public area outside the job site and observe the work before entering the job site and announcing their visit. After entering the job site, the compliance officer is required to meet with the appropriate employer representative. This could be the project manager, superintendent, foreman, or supervisor on the site.

The compliance officer is required to show a picture identification card from the U.S. Department of Labor. State inspectors are required to display similar credentials. A note must be made of the compliance officer's name, serial number, and base. The officer generally will provide a business card.

If for any reason the identification is suspicious, the information should be verified with the office that the officer claims as a base. People have been known to use falsified government identification to enter construction sites. Information collected during bogus inspections can be used by thieves. For example, a thief may want to identify what tools, materials, and equipment are stored on site, where they are kept, and the type of job-site security. This information can be used to plan a job-site robbery.

The manager should not be fooled by con artists. The compliance officer's credentials must be checked and verified. If the officer objects, there is a good chance that the person is a phony. Anyone who tries to collect a penalty during an inspection or promotes the sale of a product or service at any time is not a compliance officer. Suspected impostors must be reported immediately to the police.

OPENING CONFERENCE

After verifying the compliance officer's identity, the company office is notified for instructions. The chief executive officer (CEO) or safety director may want to be present during the inspection. Most compliance officers are reasonable people and will be willing to wait a little while for someone from the company office to arrive on site.

Management representatives for all contractors on site should be called together for an opening conference with the compliance officer. The opening conference allows the compliance officer to explain the purpose of the visit. At this time, it is stated whether the visit is part of the planned inspection program or if it was initiated because of a complaint. Accidents that result in a fatality or serious injury also trigger investigations and inspections.

No matter the reason for an inspection, the manager has a right to request a warrant before allowing the inspector to continue with the inspection. Top management should determine in advance if a warrant must be requested.

WARRANTS

There are mixed opinions about asking for warrants. However, only the owner or contractor in charge of the job site, such as the construction manager or general contractor, has the right to ask for a warrant before allowing the compliance officer to make the inspection. If the contractor in charge does not request a warrant, the compliance officer may proceed with the inspection. This would include all construction operations in progress and all equipment and tools on site.

Some people believe that if they ask for a warrant it will only make the compliance officer angry because the inspection will have to be postponed until they can obtain a warrant from a federal court. This would require the compliance officer to prove to a federal judge that the reason for the inspection was based on OSHA's random selection process, imminent danger (life-or-death situation), an employee complaint, a fatality, or hospitalization of three or more employees due to an accident on the job. This could take an hour or two, or days.

Others believe that a warrant should be requested for all OSHA inspections, because it is their right to ask for a warrant and they want to know for sure that their job site was selected for inspection for legitimate reasons and not because the compliance officer had nothing better to do or singled them out based on a hidden agenda. Most compliance officers are reasonable people who will not be offended if a warrant is requested, because it is the company's constitutional right under the OSHA act. However, according to some people, some compliance officers will take offense, return with a warrant, inspect the job site, and "get even."

Note: OSHA officers can stand in public areas where they can observe construction operations without a warrant. Pictures and observations made from public areas can be used to obtain a warrant and could also be used to issue citations after a warrant is obtained.

Asking for a warrant is a management decision that should be made prior to an inspection taking place. Every company should have an inspection policy so supervisors know what they are expected to do if a compliance officer or somebody else visits one of their job sites. If a manager is not familiar with this policy, he or she should ask management what is expected in the event of an OSHA inspection. Employees should be instructed to direct compliance officers and other visitors, such as an insurance company loss control representative or building inspector, to the supervisor in charge. Nobody should be permitted on site without permission.

FOCUSED INSPECTION

The concept of focused inspection is a significant departure from how OSHA has previously conducted construction inspections. The plan is to recognize the efforts of responsible contractors who have implemented effective safety and health programs. OSHA believes this initiative will encourage more contractors to prepare and implement effective safety and health programs.

Before touring the job site, the compliance officer will determine if the general contractor, prime contractor, or other such entity has implemented a plan to coordinate job-site safety efforts. The review would determine whether an adequate safety and health program exists and if a designated competent person is responsible for implementing the program. If both criteria are met, the compliance officer will proceed with a focused inspection. When either criterion is not met, the compliance officer will proceed with a comprehensive inspection.

When assessing the safety program, the compliance officer will consider:

- the comprehensiveness of the safety program;
- the degree to which the program has been implemented;
- the presence of competent persons as required by the relevant standards;
- how the program is enforced; and
- management policies and activities, employee involvement, and training.

If the project qualifies for a focused inspection, the officer will conduct a walk-around inspection with emphasis placed on the four leading causes of fatalities and injuries:

(1) falls (floors, platforms, roofs, pits, shafts);

(2) struck by (falling objects, vehicles, equipment);

(3) caught in/between (cave-ins, unguarded machinery, equipment); and

(4) electrical (overhead power lines, power tools and cords, outlets, temporary wiring, grounding, ground-fault).

Note: State OSHA plans have been encouraged to implement the focused inspection initiative. Not all states have embraced this concept.

Job-site Tour

After determining whether the inspection will be focused or comprehensive, the compliance officer will proceed with the inspection, accompanied by a management and employee representative. During the focused inspection, the compliance officer will talk to employees to see if they are familiar with the safety program. Any serious hazards observed during the walk-around inspection may be cited.

The focused inspection plan does not require the compliance officer to inspect the entire project if conditions observed prove that the safety and health program is in place at the job site. If conditions at the job site are poor, the compliance officer may stop the focused inspection and switch to a comprehensive inspection. The identification of a serious violation during the focused inspection does not require the compliance officer to terminate the focused inspection. Compliance officers have some latitude in deciding whether to continue with the focused inspection.

If the project does not qualify for a focused inspection, the compliance officer will make a comprehensive inspection of the entire project. The compliance officer will observe most, if not all, operations, consult with employees, take pictures, and take instrument readings when necessary.

Employees will be consulted during the inspection. The compliance officer has the right to stop and question workers in private about safety and health conditions and practices in their work place. Employees are protected under the Occupation Safety and Health Act from discrimination for exercising their safety and health rights.

During the inspection, the compliance officer will point out any unsafe conditions observed. The compliance officer's statements must be documented. It is a good practice to take pictures of the alleged hazard and make a diagram and take measurements when appropriate.

If the inspection was based on a complaint, the compliance officer's movements should be limited by taking the officer by the most direct route to the area of the alleged hazard.

Correcting Hazards

Whenever possible, the condition should be corrected immediately. Although correcting the hazard may not eliminate a citation and penalty, it demonstrates good faith, and the compliance officer will generally take this into consideration. In addition, a hazard that could injure an employee is eliminated.

Closing the Conference

After the inspection, a closing conference will be held between the compliance officer and the management representatives. During this conference, the compliance officer will discuss the unsafe or unhealthy conditions observed during the tour. The manager will be advised of the violations but will not be advised of proposed penalties at this time.

Citations and Penalties

After the compliance officer reports the findings, the area director determines what citations and proposed penalties will be issued. The citations will inform the employer of the regulations or standards that allegedly were violated. The employer will also be advised of the length of time proposed for abatement.

The citations, abatement period, and proposed penalties will be sent to the employer by certified mail. The employer is required to post the citation at or near the place a violation occurred for three days or until the violation is abated, whichever is longer.

Informal Conference

An employer has fifteen working days to file a formal notice of contest. In the interim, the employer can request an informal hearing to discuss the alleged violations, the time set for abatement, and the penalty. Employee representatives may be invited to attend the meeting. The area director is authorized to enter into a settlement agreement that revises the citations or penalties to avoid prolonged legal disputes.

Notice of Contest and Formal Hearing

If an employer is not satisfied with the result of the informal conference, the company can file another written "notice of contest" within fifteen working days of receiving the citations. The employer can contest the violation, the time set for abatement, and/or the penalty. A copy of the notice of contest must be posted in a prominent location or given to the appropriate employee representative, such as the shop steward.

After receipt of the notice of contest, the OSHA area director will forward the case to the Occupational Safety and Health Review Commission. The commission is an independent agency not associated with OSHA or the Department of Labor. The commission will assign the case to an administrative law judge (ALJ), who will schedule a hearing near the employer's work place. A hearing will be held, witnesses and testimony will be heard from both sides, and a decision will be made.

States with their own occupational safety and health programs have a state system of review and appeal. The procedures generally are similar to federal OSHAs, but cases will be reviewed by a state review board or equivalent authority.

Once the ALJ or state equivalent rules, the employer or OSHA may appeal to the commission. Commission rulings may be appealed to the appropriate U.S. Court of Appeals.

OSHA INSPECTION PROCEDURE

The following was printed with permission from the *National Utility Contractors Association's Safety Manual*. It describes step by step the procedures that can be expected to occur in an OSHA inspection.

(1) Courteously greet the compliance officer and introduce yourself.

(2) Examine officer's credentials and politely ask the reason for the visit. If the visit is a planned inspection or due to a complaint, ask if the officer has a

search warrant. Follow your company policy regarding search warrants.

(3) Advise the officer that company policy regarding OSHA inspections requires you to contact the company office. Immediately contact the company office for directions on how to proceed.

(4) The safety director may want to be present during the inspection, in which case ask the officer to wait until the safety director arrives.

(5) If the officer has a warrant and insists on proceeding with the inspection before the safety director arrives, accompany the officer. Answer all questions truthfully, but do not volunteer information the officer does not ask for. Take notes of any defects or deficiencies the officer points out. If possible, have someone correct the alleged hazards immediately. Explain items that the officer may not understand or may misinterpret, but do not argue with the officer.

(6) All records must be readily available. If requested, let the compliance officer see safety program materials, accident records, inspection records, etc.

(7) When possible, take pictures of everything the compliance officer photographs, and take additional pictures from different angles. If a camera is not available, count and make note of the pictures the officer takes. Make note of employees interviewed or questioned by the officer.

(8) If the compliance officer takes any measurements, take the same measurements yourself.

(9) Machinery or equipment not meeting the standards, should be immediately shut down. Employees exposed to hazards should be removed from the allegedly dangerous area. The safety coordinator must be notified as soon as possible of either situation.

(10) Generally, upon completion of the inspection, the compliance officer will hold a closing conference.

The officer will review the alleged violations with you. Take notes so that you can inform the safety coordinator of the results.

As always, it should be the company's policy to abide by all safety regulations to the best of its ability; therefore, cooperation in matters of this type is mandatory.

Unit 12 – Where to Get Help

Sometimes just knowing where to get information or assistance is of great value (e.g., knowing where to get a copy of the OSHA standards).

It is easy. Contact the Government Printing Office. OSHA, a company's insurance carrier, trade associations, and the local library can provide much of the information needed.

OSHA provides free information on many topics (e.g., ground fault circuit interrupters, fall protection, hazard communication, excavations, personal protective equipment, lockout/tagout). OSHA also offers free safety consultation services through state labor departments or industrial commissions.

A company's insurance carrier may have a loss control department staffed with consultants that provide free safety and health information for customers. The company's insurance agent or carrier can determine what services are available.

Local or national trade associations (e.g., NUCA, ARTBA, ASA, AGC, ABC, NAPA, NAHB, ACCA, MCA, NADC, NAPHCC, NECA, NRCA, PDCA, SMACNA, and so on) offer safety information, training programs, video tapes, and more. One phone call can open many doors.

The local library probably has books on safety, copies of state and local regulations, and other safety information. It is worth a visit.

Information is plentiful, but a company safety director has to take the time to make a call or two to find what is needed. There is no excuse for not using all these resources. In some cases, the information might be as close as the company's safety department.

The following list of OSHA area offices and state consultant's telephone numbers may be used to call for information or assistance.

70

<u>Figure 12.1</u>
Information is available from OSHA.

OSHA AREA OFFICES

Informational services, publications, audiovisual aids, and technical advice are available through OSHA area offices. The state consultants are funded by OSHA to provide free safety and health consultant and job-site inspection. Following is a list of phone numbers for each state's, plus Puerto Rico and the Virgin Islands, OSHA area office and state consultant.

Alabama

Birmingham	(205) 731-1534
Mobile	(334) 441-6131
Consultants	(205) 348-3033

Alaska

Anchorage	(907) 271-5152
Consultants	(907) 269-4939

Arizona
| Phoenix | (602) 640-2007 |
| *Consultants* | (602) 542-5795 |

Arkansas
| Little Rock | (501) 324-6291 |
| *Consultants* | (501) 682-4522 |

California
Sacramento	(916) 978-5641
San Diego	(619) 569-9071
San Francisco	(415) 744-6670
Consultants	(415) 703-4441

Colorado
Denver	(303) 844-5285
Englewood	(303) 43-4500
Consultants	(303) 491-6151

Connecticut
Bridgeport	(203) 579-5580
Hartford	(860) 240-3152
Consultants	(203) 566-4550

Delaware
| Wilmington | (302) 573-6115 |
| *Consultants* | (302) 577-3908 |

District of Columbia
| Washington | (202) 523-1452 |
| *Consultants* | (202) 576-6339 |

Florida
Fort Lauderdale	(954) 424-0242
Jacksonville	(904) 232-2895
Tampa	(813) 626-1177
Consultants	(904) 488-3044

Georgia
| Savannah | (912) 652-4393 |
| Smyrma | (770) 984-8700 |

Tucker	(404) 493-6644
Consultants	(404) 894-2646

Guam

Consultants	(671) 647-4202

Hawaii

Honolulu	(808) 541-2685
Consultants	(808) 586-9116

Idaho

Boise	(208) 334-1867
Consultants	(208) 385-3283

Illinois

Calumet City	(708) 891-3800
Des Plaines	(847) 803-4800
Fairview Heights	(618) 632-8612
North Aurora	(708) 896-8700
Peoria	(309) 671-7033
Consultants	(312) 814-2337

Indiana

Indianapolis	(317) 226-7290
Consultants	(317) 232-2688

Iowa

Des Moines	(515) 284-4794
Consultants	(515) 281-5352

Kansas

Overland Park	(913) 236-3220
Wichita	(316) 269-6644
Consultants	(913) 296-4386

Kentucky

Frankfort	(502) 227-7024
Consultants	(502) 564-6895

Louisiana

Baton Rouge	(504) 389-0474
Consultants	(504) 342-9601

Maine

Augusta	(207) 622-8417
Bangor	(207) 941-8177
Consultants	(207) 624-6460

Maryland

Baltimore	(410) 962-2840
Consultants	(410) 333-4218

Massachusetts

Braintree	(617) 565-6924
Methuen	(617) 565-8110
Springfield	(413) 785-0123
Consultants	(617) 969-7177

Michigan

Lansing	(517) 377-1892
Consultants	(517) 332-1809

Minnesota

Minneapolis	(612) 348-1994
Consultants	(612) 297-2393

Mississippi

Jackson	(601) 965-4606
Consultants	(601) 987-3981

Missouri

Kansas City	(816) 483-9531
St. Louis	(314) 425-4249
Consultants	(314) 751-3403

Montana

Billings	(406) 247-7494
Consultants	(406) 444-6418

Nebraska

Omaha	(402) 221-3182
Consultants	(402) 471-4717

Nevada

Carson City	(702) 885-6963
Consultants	(702) 486-5016

New Hampshire

Concord	(603) 225-1629
Consultants	(603) 271-2024

New Jersey

Avenel	(908) 750-3270
Hasbrouck	(201) 288-1700
Marlton	(609) 757-5181
Parsippany	(201) 263-1003
Consultants	(609) 292-3923

New Mexico

Albuquerque	(505) 248-5302
Consultants	(505) 827-2877

New York

Albany	(518) 464-6742
Bayside	(718) 279-9060
Bowmanvilles	(716) 684-3891
New York	(212) 466-2481
Syracuse	(315) 451-0808
Tarrytown	(914) 524-7510
Westbury	(516) 334-3344
Consultants	(518) 457-2481

North Carolina

Raleigh	(919) 856-4770
Consultants	(919) 662-4651

North Dakota

Bismarck	(701) 250-4521
Consultants	(701) 328-5188

Ohio

Cincinnati	(513) 841-4132
Cleveland	(216) 522-3818
Columbus	(614) 469-5582
Toledo	(419) 259-7542
Consultants	(614) 644-2631

Oklahoma

Oklahoma City	(405) 231-5351
Consultants	(405) 528-1500

Oregon

Portland	(503) 326-2251
Consultants	(503) 378-3272

Pennsylvania

Allentown	(610) 776-0592
Erie	(814) 833-5758
Harrisburg	(717) 782-3902
Philadelphia	(215) 597-4955
Pittsburgh	(412) 644-2903
Wilkes-Barre	(717) 826-6538
Consultants	(412) 357-2396

Puerto Rico

Hato Rey	(787) 277-1560
Consultants	(809) 754-2171

Rhode Island

Providence	(401) 528-4669
Consultants	(401) 277-2438

South Carolina

Columbia	(803) 765-5904
Consultants	(803) 734-9599

South Dakota

Consultants	(605) 688-4101

Tennessee

Nashville	(615) 781-5423
Consultants	(615) 741-7036

Texas

Austin	(512) 482-5783
Corpus Christi	(512) 888-3420
Dallas	(214) 320-2400
El Paso	(915) 534-7004
Fort Worth	(817) 428-2470
Houston South	(713) 286-0583
Houston North	(713) 591-2438
Lubbock	(806) 743-7681
Consultants	(512) 440-3834

Utah

Salt Lake City	(801) 487-0680
Consultants	(801) 530-6868

Vermont
   Consultants   (802) 828-2765
Virgin Island
   Consultants   (809) 772-1315
Virginia
   Norfolk   (804) 441-3820
   Consultants   (804) 786-8707

Washington
   Bellevue   (360) 553-7520
   Consultants   (206) 956-5443
West Virginia
   Charleston   (304) 347-5937
   Consultants   (303) 558-7890
Wisconsin
   Appleton   (414) 734-4521
   Eau Claire   (715) 832-9019
   Madison   (608) 264-5388
   Milwaukee   (414) 297-3315
   Consultants   (414) 521-5188
Wyoming
   Consultants   (307) 777-7786

PRINTING OFFICES

OSHA Publications Office   (202) 219-4667
Free copies of many OSHA documents are available, including recent final rules, proposed rules, brochures, forms, and so on.

OSHA Government Printing Office   (202) 512-1800
OSHA and other government documents, such as the *OSHA Construction Standards 29 CFR 1926* and quantities of pamphlets, may be purchased through this office.

OSHA Training Institute   (847) 297-4810

APPROVED STATE OSHA SAFETY AND HEALTH PROGRAMS

States with approved plans operate independent of federal OSHA. Inspections are performed by state inspectors. Citations and penalties are issued by the state labor departments or industrial commissions. For information contact the following:

Alaska Department of Labor	(907) 465-2700
Industrial Commission of Arizona	(602) 542-5795
California Department of Industrial Relations	(415) 703-4590
Connecticut Department of Labor	(203) 566-5123
Hawaii Department of Labor	(808) 586-8844
Indiana Department of Labor	(317) 232-2378
Iowa Division of Labor	(515) 281-3447
Kentucky Labor Cabinet	(502) 564-3070
Maryland Division of Labor and Industry	(410) 333-4179
Michigan Department of Labor	(517) 373-9600
Minnesota Department of Labor and Industry	(612) 296-2342
Nevada Division of Industrial Relations	(702) 687-3032
New Mexico Environmental Department	(505) 827-2850
New York Department of Labor (municipal)	(518) 457-2741
North Carolina Department of Labor	(919) 662-4585
Oregon Occupational Safety and Health Division	(503) 378-3272
Puerto Rico Department of Labor	(809) 754-2119
South Carolina Department of Labor	(803) 734-9594
Tennessee Department of Labor	(615) 741-6880
Vermont Department of Labor and Industry	(802) 828-2288
Virgin Islands Department of Labor	(809) 773-1994
Virginia Department of Labor and Industry	(804) 786-9873
Washington Department of Labor	(360) 902-4200
Wyoming Occupational Safety and Health	(307) 777-7786

OTHER SOURCES

National Institute of Occupational Safety and Health

NIOSH Publications	(513) 533-8287
NIOSH Inquiries	(800) 35-NIOSH

Department of Transportation (DOT)   (202) 366-5580

Federal Highway Administration
(FHWA)   (202) 366-0650
Information and requirements for commercial drivers
licensing (CDL) .

Research and Special Programs
Administration   (202) 366-4433
Information and requirements for hazardous materials.
Hazardous Materials Hotline   (800) 424-9158

Environmental Protection Agency (EPA)   (202) 382-2080

National Safety Council
The NSC offers books, educational materials, pamphlets,
video tapes and training programs for sale in addition to
other services.

Central Region	(800) 621-7619
Western Region	(800) 848-5588
Northeastern Region	(800) 432-5251
Southeastern Region	(800) 441-5103

Mine Safety & Health Administration   (304) 547-2031

Unit 13 – Making Inspections

PURPOSE OF INSPECTIONS

One of the most effective ways to identify hazards is to conduct safety and health inspections. Properly conducted inspections improve operations and increase efficiency, effectiveness, and profitability.

The inspection's purpose is to identify and correct procedures, activities, and conditions that create hazards and cause accidents. Inspections also help ensure that a company complies with applicable regulations.

As part of an employer's accident prevention responsibilities, OSHA requires employers to provide frequent and regular inspections of job sites, equipment, and materials. These inspections must be made by a competent person.

TYPES OF INSPECTIONS

There are two basic types of inspections: continuous and periodic. Inspections should be conducted by a competent person, generally the project manager, foreman, or supervisor.

Continuous inspections are informal with no set schedule, plan, or checklist. They are conducted daily. The individual making the inspection must be familiar with the workers, equipment, materials, the job at hand, and applicable regulations. The continuous inspection is conducted as part of the competent person's routine job-site tour.

As the job site is toured each day to determine how work is progressing, the supervisor should always be on the lookout for safety and health hazards. Identified hazards should be corrected immediately, or workers should be removed from exposure until the hazard can be controlled or eliminated.

For example, if a worker is observed standing on top of a ladder (an unsafe act), the worker should be told to get down and get a taller ladder. If workers are observed using a frayed extension cord (an unsafe condition), the cord should be removed from the work area and destroyed or tagged defective.

By taking immediate action, the hazard can be controlled or eliminated and the possibility of an accident can be reduced. The crew will realize that unsafe actions or conditions are not tolerated. They will also see that their supervisor is constantly looking out for them and is serious about safety.

Periodic inspections are formal inspections that are planned and scheduled. These inspections are deliberate, thorough, and systematic. Periodic inspections include complete job-site audits and inspection of specific equipment or operations. These inspections may be scheduled daily, weekly, monthly, quarterly, semi-annually, or annually.

Periodic inspections often are used to ensure that the job site is thoroughly inspected and to comply with safety and health regulations. For example, OSHA requires employers to inspect cranes daily. OSHA also requires cranes to be thoroughly inspected at least annually.

WHO SHOULD MAKE INSPECTIONS?

Management is responsible for ensuring that inspections are performed. Inspections should be completed by individuals or teams that are familiar with the operations and equipment to be inspected.

Some inspections can be delegated to employees who operate and maintain equipment or tools. For example, equipment operators are responsible for inspecting their equipment each day before starting work. Delegating responsibility does not eliminate management's responsibility for ensuring that inspections are completed.

Inspectors should possess the following qualifications:

- knowledge of the work areas and work being inspected;
- knowledge of the materials, tools, and equipment in use;
- knowledge of relevant standards, regulations, and codes;

- knowledge of procedures for evaluating and reporting hazards; and
- knowledge of methods used to ensure corrective action and followup.

Individuals responsible for inspections should determine the following before making an inspection:

- What needs to be inspected?
- What aspects of the operation or equipment must be inspected?
- What conditions should be looked for or observed?
- What are the established safe procedures for operating equipment, machines, and tools that will be found during inspections?
- How often must inspections be made?

WHAT MUST BE INSPECTED?

All operations, equipment, tools, and materials should be inspected in accordance with the procedures set forth in the company's safety plan. Following are some things that should be inspected when applicable:

- housekeeping and sanitary conditions
- environmental conditions
- first-aid supplies, eyewash stations, and safety showers
- fire protection equipment
- walkways and roadways
- fall protection
- equipment and machinery
- electrical equipment and power sources
- hand and power tools
- scaffolding
- exits and exit paths
- hoisting and rigging equipment
- storage areas
- personal protective equipment
- materials handling equipment

- trench shoring equipment
- ladders and stairways
- hazardous materials and supplies
- motor vehicles
- warning signs and barricades
- storage of flammable gases and liquids
- welding and cutting equipment

Particular attention should be paid to the following:

- safety devices
- guards
- controls
- point of operation
- electrical equipment and installations
- mechanical equipment
- procedures

When making inspections, observe worker behavior to ensure they are following safe practices and procedures. Watch for:

- violations of safe procedures
- unauthorized use of tools or equipment
- removal of guards or other safety devices
- creation of unsafe or unsanitary conditions
- operation of equipment at unsafe speeds
- defective tools or equipment
- improper use of tools or equipment
- overloaded cranes, hoists, or other equipment
- failure to implement lockout procedures
- unauthorized entry into restricted areas
- other, depending on the work in progress

INSPECTION FREQUENCY

Inspection frequency is established based on the type of operation, the potential for an accident, the type of hazards that could exist, the level of exposure, how quickly the work could become unsafe, and applicable regulations.

For example, consider a trenching operation to install a new water line. The potential for an accident is high unless a protective system is used. The primary hazard is a cave-in. Employee exposure to an unprotected area could result in death or serious injury. Conditions can change quickly as the trenching operation moves along. OSHA requires that all excavations must be inspected daily and as often as necessary to ensure worker safety.

In this case, inspections must be continuous and ongoing daily. Therefore, inspections must be made as often as necessary to ensure the safety of workers and the general public.

DOCUMENTING INSPECTIONS

Legal opinions are mixed about documenting inspections. Some experts believe inspections should be thoroughly documented to ensure that nothing is overlooked. Other experts say that inspection documentation can be used against a company in a legal battle. In some situations, documentation is mandatory. Therefore, company policy should be followed regarding documentation of inspections.

If an inspection checklist is used, it may not include everything that needs to be inspected. Inspection reports or checklists should include:

- project name
- area or equipment inspected
- date of inspection
- time of inspection
- name and signature of person making inspection
- location of hazards
- description of hazards
- identity of hazardous machine or equipment
- recommended corrective action
- remarks

General Inspection Form

Location _____

_____

Date of Inspection _____

This list is intended only as a reminder. Look for other potentially unsafe acts or conditions, and then report them so that corrective action can be taken. Note particularly whether unsafe acts or conditions that have caused accidents have been corrected. Note also whether potential accident causes identified as unsatisfactory on previous inspections have been corrected. All items that are found to be unsatisfactory should be listed. Mark "NA" if not applicable under the "No" column. Do not leave any blanks because of legal and insurance requirements.

		Yes	No
1.0	**Fire Protection**		
1.1	Extinguishers charged and accessible?	___	___
1.2	If available, standpipes, hoses, sprinkler heads and valves in good condition and accessible?	___	___
1.3	Stairs clear and in good condition?	___	___
1.4	Hollow pan stairways filled?	___	___
1.5	Exits and exit paths clearly marked?	___	___
1.6	Flammables properly stored? (gasoline, paint solvents, acetylene, propane tanks, and so on)	___	___
1.7	Evacuation plan as required by OSHA available?	___	___
2.0	**Housekeeping**		
2.1	Aisles, stairs, and floor free of obstructions?	___	___
2.2	Materials and supplies stored and piled in designated areas?	___	___
2.3	Regular removal of trash and debris?	___	___
2.4	Are all work areas lighted?	___	___
2.5	Work areas neat and orderly?	___	___
3.0	**Tools**		
3.1	Tool casings in good condition?	___	___
3.2	Wiring for all power tools in good condition?	___	___
3.3	Electric tools grounded (unless double insulated)?	___	___
3.4	Extension cords grounded and in good condition?	___	___
3.5	Hand tools in good condition?	___	___
3.6	Tools stored in designated location?	___	___
3.7	Ladders free of cracks and damage?	___	___

4.0	**Personal Protective Equipment (PPE) when required**	**Yes**	**No**
	4.1 Goggles, safety glasses, face shields used?	___	___
	4.2 Gloves used?	___	___
	4.3 Rubber boots worn?	___	___
	4.4 Hard hats worn?	___	___
	4.5 Employees wearing proper work shoes? (no sneakers or open toe shoes)	___	___
	4.6 Other PPE required, available, and used?	___	___
5.0	**Material Handling Equipment**		
	5.1 Carts in good condition?	___	___
	5.2 Cart wheels free and rolling smoothly?	___	___
	5.3 Hoist opening equipped with removable railing?	___	___
	5.4 Hoist cables and hooks inspected?	___	___
	5.5 Materials securely stacked?	___	___
	5.6 Employees trained and certified to operate equipment?	___	___
6.0	**Machinery**		
	6.1 Point of operation guards in place? (i.e., blade guards on some)	___	___
	6.2 Pulley belt assemblies guarded?	___	___
	6.3 Gear assemblies guarded?	___	___
	6.4 Shafts guarded?	___	___
	6.5 Are there any oil leaks?	___	___
	6.6 Two hand controls working properly?	___	___
	6.7 Is electric wiring in good condition? (not frayed, not pulled out of connectors, and so on)	___	___
	6.8 Lockout policy and tag procedure used?	___	___
7.0	**Welding Equipment and Operations**		
	7.1 Oxygen and acetylene welding equipment with flash arrestors?	___	___
	7.2 Compressed gas cylinders secured and capped when in storage?	___	___
	7.3 Cylinders mounted on a cart or secured in an upright position when in use?	___	___
	7.4 Is oxygen separated from flammables and combustibles by at least 20 ft or a 5-ft high noncombustible wall when in storage?	___	___
	7.5 Gas hoses and gauges in good condition?	___	___
	7.6 Proper eye protection available and used?	___	___
8.0	**Electrical**		
	8.1 All circuits used to provide power to tools and equipment protected by ground fault circuit interrupter?	___	___
	8.2 Assured equipment or grounding conductor program in effect?	___	___

9.0 First-Aid and Emergency Procedures

    9.1 First-aid kits available and properly stocked?   ___ ___

    9..2 Are there any trained first-aid personnel available ___ ___
        on all shifts?

    9.3 Emergency phone numbers posted?   ___ ___

10.0 Unsafe Practices

    10.1 Employees moving powered equipment at   ___ ___
        proper speed, not excessive?

    10.2 Do employees use the proper lifting technique? ___ ___

    10.3 Do employees operate machinery with guards   ___ ___
        in place?

    10.4 All employees wearing proper personal   ___ ___
        protective equipment for the task performed?

    10.5 Confined space procedures followed?   ___ ___

11.0 Employee Training

    11.1 All employees received hazard identification   ___ ___
        training?

    11.2 All employees trained in hazard communication? ___ ___

    11.3 All employees trained in appropriate fire fighting ___ ___
        response?

    11.4 All employees trained in evacuation procedures? ___ ___

    11.5 Appropriate employees must be trained in   ___ ___
        lockout/tagout procedures.

    11.6 Confined space training for appropriate   ___ ___
        employees?

    11.7 Foreperson holding tool box talks?   ___ ___
        (weekly, monthly—circle one)

12.0 Other Specific Job-Site Safety Requirements

_____

_____ ___ ___

_____ ___ ___

I understand that lack of inspections as well as falsification of inspection forms is a violation of company procedure and may be a civil and/or criminal violation of federal and/or state laws and/or regulations.

_____   _____

Responsible Management Position   Date

<u>Figure 13.1</u>

Follow company policy when documenting inspections.

Corrective Action

Identified hazards should be corrected as soon as possible. Where possible, immediate corrective action should be taken to eliminate or control a hazard. Document corrective action taken.

Defective equipment should be removed immediately from service and tagged "Defective—Do Not Use." If a worker is not following safety procedures, immediate action must be taken to ensure that the worker knows and follows proper procedures. Disciplinary action must be taken if the worker continues to violate procedures. If workers are exposed to a hazard that cannot be corrected immediately, they should be removed from the area until the hazard is corrected. When corrective action is not taken quickly to control hazards, accidents will occur.

Unit 14 – Recordkeeping

OSHA RECORDS

The Occupational Safety and Health Administration requires employees to keep records of work-related injuries and illnesses. An injury or illness is work related if an event or exposure in the work environment either caused or contributed to an employee injury or illness.

Occupational injury is any injury, such as a cut, fracture, sprain, amputation, eye injury, and so on, which results from a work-related accident or from an exposure involving a single incident in the work environment.

Occupation illness is any abnormal health condition or physical disorder, other than one resulting from an occupational injury, caused by work-related exposure to environmental factors. It includes acute and chronic illnesses or diseases that may be caused by:

- inhalation of chemicals, fumes, or vapors;
- absorption of chemicals, hazardous materials, or radiation through the skin;
- ingestion of chemicals or hazardous materials; or
- contact with chemicals, hazardous materials, or hot or cold environments

These acute and chronic illnesses and diseases may result in:

- skin disease or disorder,
- disease of the lungs,
- respiratory conditions,
- poisoning,
- disorders related to temperature extremes,
- disorders associated with repeated trauma, and
- all other occupational illness.

A recordable injury or illness is one that meets the following criteria:

(1) an injury or illness exists; and

(2) is work related; and

(3) results in death, loss of consciousness, medical treatment other than first aid, or at least one day away from work, restricted work activity, or job transfer.

OSHA injury and illness logs must be maintained on an annual basis. They must be updated weekly to ensure compliance.

OSHA also requires employers to maintain an incident record for each reportable injury or illness. Only recordable injuries and illnesses must be entered on the log; however, it is a good practice to maintain records for all work-related injuries or illnesses, no matter how minor.

States with state safety plans have similar recordkeeping requirements. Recordkeeping forms may be obtained from OSHA or state plan administrator, depending on where your company is located.

ACCIDENT REPORTS

Accident reporting procedures should be included in a company's safety and health program. If they are not included, check with the company safety director to determine reporting responsibilities.

WORKER'S COMPENSATION RECORDS

Each state has its own worker's compensation laws and recordkeeping requirements that must be met before compensation payments will be paid to injured or ill workers. To receive worker's compensation, the injury or illness must be work related and the proper forms must be filed with the insurance carrier. A company could be penalized if it fails to report reportable work-related injuries or illnesses.

Although, it is unlikely that the manager will be responsible for filing the worker's compensation reports with the insurance carrier, the manager is still responsible for reporting all injuries and illnesses to a company's safety or insurance manager. They will instruct the manager in what to do and what information is needed.

TRAINING RECORDS

Training records should be maintained for all training sessions. These records should document who was trained and what training was provided. All indoctrinations, training sessions, and toolbox (tailgate) talks should be documented. Maintain a list of who attended each session and document the topic of the meeting.

The manager, foreman, or supervisor will occasionally provide individual workers with special training to show them how to perform a specific task, or operate tools or equipment. The importance of keeping records when individualized training is provided is often overlooked. If the training has been provided, the manager must take another minute to make a record of it. Even a brief notation in the job-site log or a personal notebook can be important in a hearing or court.

Unit 15 – Is More Safety and Health Training Needed?

TRAINING IS IMPORTANT TO ONE'S CAREER

There is more to job-site safety than just handing out hard hats and telling workers to wear them.

The manager should take advantage of any work-related safety training that is available: seminars, videos, manufacturer's instructions, trade journals, books, and, above all, become knowledgeable about work-related safety and health issues.

There is nothing more important than job-site safety and the safety of the crew. Responsibility for managing a crew's safety is one of great importance.

WHAT TYPE OF TRAINING IS NEEDED?

Each individual's training needs are different. The manager should consult with the company's safety director or a safety consultant to evaluate the type of training needed based on the type of work being supervised. For example, when supervising trenching and excavation operations, the supervisor should attend a competent person-training program. When supervising a job site where cranes are used, crane safety should be learned. When new tools or equipment is brought on site, the instruction manual must be read to find out how to safely operate the equipment. Hazards associated with each task must be identified and others must be instructed in a safe way to perform the task.

If the manager has not attended an OSHA 10-hour safety program or equivalent, it is a good place to start. In fact, the manager should attend a program of this type at least every few years to stay familiar with the applicable regulations.

OSHA programs often are available at a reasonable price through local trade associations and private seminars.

WHERE TO FIND TRAINING

Training is available from:

- OSHA Institute
- safety and health consultants
- National Safety Council
- American Society of Safety Engineers
- Center for Construction Education and Research
- local safety councils
- trade associations
- OSHA safety consultants
- trade unions
- insurance companies
- manufacturers
- tool and equipment distributors
- video training tapes can be purchased, rented, or borrowed. They are also available in some libraries.
- Computer-based training is becoming more readily available. It is only a matter of time before it is available over the Internet.

SAFETY TRAINED SUPERVISOR—CONSTRUCTION

Anyone with two years of construction experience, at least one year as a foreman, crew chief, or superintendent, and thirty hours of safety training who wants to achieve a recognized certification in construction safety can take a certification exam offered by the Board of Certified Safety Professionals (BCSP). The BCSP offers a certification for Safety Trained Supervisor—Construction. The examination targets job-site safety. Typical safety tasks included on the examination are conducting new employee safety orientation, performing basic hazard analysis, recognition and correction, issuing and monitoring the use of personal protective equipment, conducting safety meetings, performing hazard prevention analysis and work preplanning, inspecting tools and equipment, applying the OSHA hazard communication standard, enforcing safety standards on job sites, participating in job-site safety inspections, and responding to accidents. For information contact the BCSP, 208 Burwash Avenue, Savoy, IL 61874-9904.

Unit 16 – OSHA in a Nutshell

The following is a summary of the *OSHA 1926/1910 Construction Standards*. Although it does not include every specific requirement, it does include many of the more frequently violated construction standards.

All construction field managers (project managers, super-intendents, foremen, supervisors) should review the sections that apply to the work performed by their crews. This section should in no way be considered as a complete substitute for any provisions or the Occupational Safety and Health Act of 1970. The requirements contained herein are summarized and abbreviated. Additional and more detailed information can be obtained by referencing the actual *OSHA 1926 Construction Standards* and *1910 General Industry Standards*.

1. Abrasive Grinding
   a. All abrasive wheel bench and stand grinders shall be provided with safety guards that cover the spindle ends, nut and flange projections, and are strong enough to withstand the effects of a bursting wheel. [1926.303(b) & (c)(1)]
   b. An adjustable work rest of rigid construction shall be used on floor and bench-mounted grinders, with the work rest kept adjusted to a clearance not to exceed $1/8$ inch between rest and the surface of the wheel. [1926.303(c)(2)]
   c. All abrasive wheels shall be closely inspected and ring-tested before mounting to ensure that they are free from cracks or other defects. [1926.303(c)(7)]

2. Access to Medical and Exposure Records
   a. Each employer shall permit employees, their designated representatives, and OSHA direct access to

employer-maintained exposure and medical records. The standard limits access only to those employees who are, have been (including former employees), or will be exposed to toxic substances or harmful physical agents. [1910.20(a) & (b)(3)]

b. Each employer must preserve and maintain accurate medical and exposure records for each employee. Exposure records and data analyses based on them are to be kept for thirty years. Medical records are to be kept for at least the duration of employment plus thirty years. Background data for exposure records, such as laboratory reports and work sheets need to be kept for only one year. Records of employees who have worked for less than one year need not be retained after employment, but the employer must provide these records to the employee upon termination of employment. First-aid treatment need not be retained for any specified period. [1910.20(d)]

3. Accident Recordkeeping and Reporting Requirements

a. Each employer shall maintain in each establishment an OSHA log and summary of all recordable injuries and illnesses (resulting in a fatality, hospitalization, lost workdays, medical treatment, job transfer or termination, or loss of consciousness) for that establishment, and enter each recordable event no later than six working days after receiving the information. Where the complete log and summary records are maintained at a place other than the establishment, a copy of the log shall be available at the establishment, which reflects separately the injury and illness experience of that establishment complete and current to a date within forty-five calendar days. [1904.2(a) & (b)(2)]

b. In addition to the log of occupational injuries and illnesses, each employer shall have available for inspection at each establishment within six working days after notification of a recordable case, an incident report for each occupational injury or illness for that establishment. [1904.4]

c. Each employer shall post an annual summary of occupational injuries and illnesses for each establishment, compiled from the collected OSHA log and summary, including the year's totals, calendar year covered, company name, establishment name and address, certification signature, title, and date. [1904.5]

d. The log and summary, the incident reports, and the annual summary shall be retained in each establishment for five years following the end of the year to which they relate. Records shall be made available, as authorized, upon request. [1904.6 & 1904.7(a) & (b)]

e. Within eight hours after its occurrence, an employment accident that is fatal to one or more employees or results in the hospitalization of three or more employees shall be reported by the employer, either orally or in writing, to the nearest OSHA area office or the national 1-800-321-OSHA. [1904.8]

4. Air Tools

a. Pneumatic power tools shall be secured to the hose in a positive manner to prevent accidental disconnection. [1926.302(b)(1)]

b. Safety clips or retainers shall be securely installed and maintained on pneumatic impact tools to prevent attachments from being accidentally expelled. [1926.302(b)(2)]

c. The manufacturer's safe operating pressure for all fittings shall not be exceeded. [1926.302(b)(5)]

d. All hoses exceeding ½ inch inside diameter shall have a safety device at the source of supply or branch line to reduce pressure in case of hose failure. [1926.302(b)(7)]

5. Asbestos

   a. The employer shall ensure that no employee is exposed to an airborne concentration of asbestos in excess of 0.1 fiber per cubic centimeter of air as an eight-hour time-weighted average (TWA) and the excursion limit of 1.0 fiber per cubic centimeter of air (1 f/cc) as averaged over a sampling period of thirty minutes. [1926.1101(c)(1) & (c)(2)]

   b. An employer performing work requiring the establishment of a regulated area shall inform other employees on site of the nature of the work. Asbestos hazards shall be abated by the contractor who created or controls the source of asbestos contamination. All employers of employees exposed to the hazard shall comply with applicable protective provisions to protect their employees. [1926.1101(d)(1) through (d)(3)]

   c. All Class I, II, and III asbestos work shall be conducted within regulated areas. All other operations covered by the standard shall be conducted within a regulated area where airborne concentrations of asbestos exceed, or there is a reasonable possibility they may exceed, PEL. [1926.1101(e)(1)]

   d. Each employer who has a workplace or work operation covered by this standard shall perform monitoring to determine accurately the airborne concentrations of asbestos to which employees may be exposed. [1926.1101(f)(1)(i)]

   e. Respirators must be used (1) while feasible engineering and work practice controls are being installed or implemented; (2) during maintenance and repair activities or other activities where engineering and work practice controls are insufficient to reduce employee exposure; and (4) in emergencies. [1926.1101(h)(1)(i) through (h)(1)(viii)]

f. The employer shall provide and require the use of protective clothing, such as coveralls or similar whole body clothing, head coverings, gloves, and foot coverings, for any employee exposed to airborne concentrations of asbestos that exceed the TWA and excursion limit. [1926.1101(i)(1)]

g. The employer shall establish a decontamination area that is adjacent and connected to the regulated area for the decontamination of employees working in the regulated area. [1926.1101(j)(1)(i)]

h. The employer shall post warning signs that demarcate the regulated areas. [1926.1101(k)(6)(i)]

i. The employer shall institute a training program for all employees who install or remove asbestos products. [1926.1101(k)(9)(i)]

j. The employer shall institute a medical surveillance program for all employees engaged in work involving levels of asbestos, at or above the action level and/or excursion limit for thirty or more days per year, or who are required by the standard to wear negative pressure respirators. [1926.1101(m)(1)(i)]

6. Belt Sanding Machines

a. Belt sanding machines shall be provided with guards at each nip point where the sanding belt runs onto a pulley. [1926.304(f)]

b. The unused run of the sanding belt shall be guarded against accidental contact. [1926.304(f)]

7. Chains (See Wire Ropes, Chains, Hooks, and so on)

8. Compressed Air, Use of

a. Compressed air used for cleaning purposes shall not exceed 30 psi and then only with effective chip guarding and personal protective equipment. [1926.302(b)(4)]

b. This requirement does not apply to concrete form, mill scale, and similar cleaning operations. [1926.302(b)(4)]

9. Compressed Gas Cylinders

   a. Valve protection caps shall be in place when compressed gas cylinders are transported, moved, or stored. [1926.350(a)(1)]

   b. Cylinder valves shall be closed when work is finished and when cylinders are empty or are moved. [1926.350(a)(8)]

   c. Compressed gas cylinders shall be secured in an upright position at all times, except if necessary for short periods of time when cylinders are actually being hoisted or carried. [1926.350(a)(9)]

   d. Cylinders shall be kept at safe distance or shielded from welding or cutting operations. Cylinders shall be placed where they cannot become part of an electrical circuit. [1926.350(b)(1) & (b)(2)]

   e. Oxygen and fuel gas regulators shall be in proper working order while in use. [1926.350(h)]

   f. For additional details not covered in this subpart, applicable technical portions of American National Standards Institute, Z-49.1-1967, Safety in Welding and Cutting, shall apply. [1926.350(j)]

   g. Oxygen and fuel gas cylinders in storage shall be separated by a 5-ft noncombustible wall or a 20-ft separation. [1926.350(j)]

   h. For additional details not covered in this subpart, applicable technical portions of American National Standards Institute (ANSI) Z49.1-1967, Safety in Welding and Cutting, shall apply. [1926.350(j)]

10. Concrete and Masonry Construction

   a. No construction loads shall be placed on a concrete structure or portion of a concrete structure unless the employer determines, based on information received from a person who is qualified in structural design, that the structure or portion of the structure is capable of supporting the loads. [1926.701(a)]

b. All protruding reinforcing steel, onto and into which employees could fall, shall be guarded to eliminate the hazard of impalement. It should be understood that the little plastic caps commonly used on rebar do **not** prevent impalement. [1926.701(b)]

c. No employee shall be permitted to work under concrete buckets while buckets are being elevated or lowered into position. [1926.701(e)(1)]

d. To the extend practical, elevated concrete buckets shall be routed so that no employee, or the fewest number of employees, are exposed to the hazards associated with falling concrete buckets. [1926.701(e)(2)]

e. Formwork shall be designed, fabricated, erected, supported, braced, and maintained so that it will be capable of supporting without failure all vertical and lateral loads that may reasonably be anticipated to be applied to the formwork. [1926.703(a)(1)]

f. Forms and shores (except those used for slabs on grade and slip forms) shall not be removed until the employer determines that the concrete has gained sufficient strength to support its weight and superimposed loads. Such determination shall be based on compliance with one of the following:

   • The plans and specifications stipulating conditions for removal of forms and shores, and such conditions have been followed; or

   • The concrete has been properly tested with an appropriate American Society for Testing Materials (ASTM) standard test method designed to indicate the concrete compressive strength, and the test results indicate that the concrete has gained sufficient strength to support its weight and superimposed loads. (ASTM, 1916 Race Street, Philadelphia, PA 19103. Telephone: (216) 299-5400). [1926.703(e)(1)]

g. A limited access zone shall be established whenever a masonry wall is being constructed. The limited access zone shall conform to the following:

- The limited access zone shall be established prior to the start of construction of the wall;

- The limited access zone shall be equal to the height of the wall to be constructed plus 4 ft, and shall run the entire length of the wall;

- The limited access zone shall be established on the side of the wall that will be unscaffolded;

- The limited access zone shall be restricted to entry by employees actively engaged in constructing the wall. No other employee shall be permitted to enter the zone; and

- The limited access zone shall remain in place until the wall adequately is supported to prevent overturning and to prevent collapse, where the height of a wall is over 8 ft, the limited access zone shall remain in place until the requirements of paragraph (h) of this section have been met. [1926.706(a)(1) through (a)(5)]

h. All masonry walls over 8 ft in height shall be adequately braced to prevent overturning and to prevent collapse unless the wall is adequately supported so that it will not overturn or collapse. The bracing shall remain in place until permanent supporting elements of the structure are in place. [1926.706(b)]

i. Lift-slab construction.

- Lift-slab operations shall be designed and planned by a registered professional engineer who has experience in lift-slab construction. Such plans and designs shall be implemented by the employer and shall include detailed instructions and sketches indicating the prescribed method of erection. [1926.705(a)]

- Jacking equipment shall be capable of supporting at least two and one-half times the load being lifted during jacking operations and the equipment shall not be overloaded. [1926.705(d)]

- No employee, except those essential to the jacking operations, shall be permitted in the building and structure while any jacking operation is taking place unless the building and structure has been reinforced sufficiently to ensure its integrity during erection. [1926.705(k)(1)]

- Equipment shall be designed and installed so that the lifting rods cannot slip out of position or the employer shall institute other measures, such as the use of locking or blocking devices, which will provide positive connection between the lifting rods and attachments and will prevent components from disengaging during lifting operations. [1926.705(p)]

11. Cranes and Derricks
   a. The employer shall comply with the manufacturer's specifications and limitations. [1926.550(a)(1)]
   b. Rated load capacities, recommended operating speeds, and special hazard warnings or instructions shall be conspicuously posted on all equipment. Instructions or warnings shall be visible from the operator's station. [1926.550(a)(2)]
   c. Equipment shall be inspected by a competent person before each use and during use, and all deficiencies corrected before further use. [1926.550(a)(5)]
   d. Accessible areas within the swing radius of the rear of the rotating superstructure shall be properly barricaded to prevent employees from being struck or crushed by the crane. [1926.550(a)(9)]
   e. Except where electrical distribution and transmission lines have been de-energized and visibly grounded at point of work, or where insulating barriers not a part

of or an attachment to the equipment or machinery have been erected to prevent physical contact with the lines, no part of a crane or its load shall be operated within 10 ft of a line rated 50 kV or below; 10 ft + 0.4 in. for each 1 kV over 50 kV for lines rated over 50 kV, or twice the length of the line insulator, but never less than 10 ft. [1926.550(a)(15)(i) & (a)(15)(ii)]

f. An annual inspection of the hoisting machinery shall be made by a competent person or by a government or private agency recognized by the U.S. Department of Labor. Records shall be kept of the dates and results of each inspection. [1926.550(a)(6)]

g. All crawler, truck, or locomotive cranes in use shall meet the requirements as prescribed in the ANSI B30.5-1968, Safety Code for Crawler, Locomotive and Truck Cranes. (ANSI, 11 West 42nd Street, New York, NY 10036. Telephone #(212) 642-4900). [1926.550(b)(2)]

h. The use of a crane or derrick to hoist employees on a personnel platform is prohibited, except when the erection, use, and dismantling of conventional means of reaching the worksite, such as a personnel hoist, ladder, stairway, aerial lift, elevating work platform, or scaffold, would be more hazardous, or is not possible because of structural design or worksite conditions.Personnel platforms must meet specific requirements. [1926.550(g)]

12. Disposal Chutes

a. Whenever materials are dropped more than 20 ft to any exterior point of a building, an enclosed chute shall be used. [1926.252(a)]

b. When debris is dropped through holes in the floor without the use of chutes, the area where the material is dropped shall be enclosed with barricades not less than 42 in. high and not less than 6 ft back from projected edges of the opening above. Warning signs of the hazard of falling material shall be posted at each level. [1926.252(b)]

13. Drinking Water

    a. An adequate supply of potable water shall be provided in all places of employment. [1926.51(a)(1)]

    b. Portable drinking water containers shall be capable of being tightly closed and be equipped with a tap. [1926.51(a)(1)]

    c. The common drinking cup is prohibited. Cup dispensers and disposable cups shall be provided. [1926.51(a)(4)]

    d. A sanitary container for unused cups and a receptacle for used cups shall be provided. [1926.51(a)(5)]

14. Electrical Installations

    a. Electrical installations made in accordance with the 1984 *National Electrical Code*® are considered to be in compliance with OSHA's electrical standards for construction, except for the following additional requirements:

    Employers must provide either ground-fault circuit interrupters (GFCIs) or an assured equipment grounding conductor program to protect employees from ground-fault hazards at construction sites. The two options follow:

    All 120-volt, single-phase, 15- and 20-ampere receptacles that are not part of the permanent wiring must be protected by GFCIs; or

    An assured equipment grounding program covering extension cords, receptacles, and cord-and plug-connected equipment must be implemented. The program must include the following:

    A written description of the program.

    At least one competent person to implement the program.

Daily visual inspections of extension cords and cord-and-plug-connected equipment for defects.

Continuity tests of the equipment grounding conductors or receptacles, extension cords, and cord-and-plug-connected equipment. These tests must be made at least every three months. [1926.402(a) & 1926.404(b)(1)(i) through (d)]

b. Lamps for general illumination must be protected from breakage, and metal shell sockets must be grounded. [1926.405(a)(2)(ii)(e)]

c. Temporary lights must not be suspended by the cords, unless they are so designed. [1926.405 (a)(2)(ii)(f)]

d. Portable lighting used in wet or conductive locations, such as tanks or boilers, must be operated at no more than 12 volts or must be protected by GFCIs. [1926.405(a)(2)(ii)(g)]

e. Extension cords must be of the three-wire type. Extension cords and flexible cords used with temporary and portable lights must be designed for hard or extra-hard usage (for example, types, S, ST, and SO). [1926.405(a)(2)(ii)(j)]

f. Listed, labeled, or certified equipment shall be installed and used in accordance with instructions included in the listing, labeling, or certification. [403(b)(2)]

15. Electrical Work Practices

a. Employers must not allow employees to work near live parts of electrical circuits, unless the employees are protected by one of the following means:

- De-energizing and grounding the parts;
- Guarding the part by insulation; or
- Any other effective means. [1926.416(a)(1)]

b. In work areas where the exact location of underground electrical power lines is unknown, employees using jack hammers, bars, or other hand tools that may contact the lines must be protected by insulating gloves, apron, or other protective clothing which will provide equivalent electrical protection. [1926.416(a)(2)]

c. Barriers or other means of guarding must be used to ensure that work space for electrical equipment will not be used as a passageway during periods when energized parts of equipment is exposed. [1926.416(b)]

d. Worn or frayed electrical cords or cables must not be used. Extension cords must not be fastened with staples, hung from nails, or suspended by wire. [1926.416(c)(1) & (c)(2)]

e. Flexible cords must be connected to devices and fittings so that strain relief is provided which will prevent pull from being directly transmitted to joints or terminal screws. [1926.405(g)(2)(iv)]

f. Equipment or circuits that are de-energized must be rendered inoperative and must have tags attached at all points where the equipment or circuits could be energized. [1926.417(b)]

16. Excavating and Trenching

a. The estimated location of utility installations, such as sewer, telephone, fuel, electric, water lines, or any other underground installations that reasonably may be expected to be encountered during excavation work, shall be determined prior to opening an excavation. [1926.651(b)]

b. Utility companies or owners shall be contacted within established or customary local response times, advised of the proposed work, and asked to establish the location of the utility underground installations prior to the start of actual excavation.

When utility companies or owners cannot respond to a request to locate underground utility installations within 24 hours (unless a longer period is required by state or local law), or cannot establish the exact location of these installations, the employer may proceed, provided the employer does so with caution, and provided detection equipment or other acceptable means to locate utility installations are used. [1926.651(b)(2)]

c. When excavation operations approach the estimated location of underground installations, the exact location of the installations shall be determined by safe and acceptable means. While the excavation is open, underground installations shall be protected, supported or removed as necessary to safeguard employees. [1926.651(b)(3) & (b)(4)]

d. Each employee in an excavation shall be protected from cave-ins by an adequate protective system except when excavations are made entirely in stable rock; or excavations are less than 5 ft (1.52 m) in depth and examination of the ground by a competent person provides no indication of a potential cave-in. [1926.652(a)(1)]

e. Protective systems shall have the capacity to resist without failure all loads that are intended or could reasonably be expected to be applied or transmitted to the system. [1926.652(a)(2)]

f. Employees shall be protected from excavated or other materials or equipment that could pose a hazard by falling or rolling into excavations. Protection shall be provided by placing and keeping such materials or equipment at least 2 ft (0.61 m) from the edge of excavations, or by the use of retaining devices that are sufficient to prevent materials or equipment from falling or rolling into excavations, or by a combination of both if necessary. [1926.652(j)(2)]

g. Daily inspections of excavations, the adjacent areas and protective systems shall be made by a competent person for evidence of a situation that could result in possible cave-ins, indications of failure of protective systems, hazardous atmospheres, or other hazardous conditions. An inspection shall be conducted by a competent person prior to the start of work and as needed throughout the shift. Inspections shall also be made after every rainstorm or other hazard increasing occurrence. These inspections are only required when employee exposure is anticipated. [1926.651(k)(1)]

h. Where a competent person finds evidence of a situation that could result in a possible cave-in, indications of failure of protective systems, hazardous atmospheres, or other hazardous conditions, exposed employees shall be removed from the hazardous area until the necessary precautions have been taken to ensure their safety. [1926.651(k)(2)]

i. A stairway, ladder, ramp or other safe means of egress shall be located in trench excavations that are 4 ft (1.22 m) or more in depth so as to require no more than 25 ft (7.82 m) of lateral travel for employees. [1926.651(c)(2)]

17. Explosives and Blasting

a. Only authorized and qualified persons shall be permitted to handle and use explosives. [1926.900(a)]

b. Explosive material shall be stored in approved facilities as required by provisions of the Internal Revenue Service regulations published in 27 CFR 181, "Commerce in Explosives." [1926.904(a)]

c. Smoking and open flames shall not be permitted within 50 ft of explosives and detonator storage magazines. [1926.904(c)]

d. Procedures that permit safe and efficient loading shall be established before loading is started. [1926.905(a)]

109

18. Eye and Face Protection

   a. Eye and face protection shall be provided when machines or operations present potential eye or face injury. [1926.102(a)(1)]

   b. Eye and face protective equipment shall meet all requirements of ANSI Z87.1-1968, "Practice for Occupational and Educational Eye and FaceProtection." [1926.102(a)(2)]

   c. Employees involved in welding operations shall be furnished with filter lenses or plates of at least the proper shade number. [1926.102(b)(1)]

   d. Employees exposed to laser beams shall be furnished suitable laser safety goggles which will protect for the specific wavelength of the laser and be optical density (OD) adequate for the energy involved. [1926.102(b)(2)]

19. Fall Protection

   a. Each employee on a walking/working surfaces (horizontal or vertical surface) with an unprotected side or edge that is 6 ft or more above a lower level shall be protected from falling with a guardrail system, safety net system, or personal fall-arrest system. Exception: When performing leading edge construction and the employer can demonstrate that it is infeasible or creates a greater danger to use these systems, the employer shall develop and implement a fall protection plan with an established and controlled access zone.) [1926.501(b)(1) & (b)(2)]

   b. Each employee shall be protected from falling through holes more than 6 ft above lower levels, by a personal fall arrest system, cover, or guardrail system. [1926.501(b)(4)]

   c. Each employee working on the face of formwork or reinforcing steel shall be protected from fall 6 ft or more by personal a fall arrest system, safety net system, or positioning device system. [1926.501(b)(5)]

d. Ramps, runways, and other walkways 6 ft or more above lower levels shall be equipped with guardrails. [1926.501(b)(6)]

e. Each employee at the edge of an excavation 6 ft or more in depth shall be protected from falling by a guardrail system, fence, or barricade when the excavations cannot be readily seen because of plant growth or other visual barrier. [1926.501(b)(7)(i)]

f. Each employee at the edge of a well, pit, shaft, and similar excavation 6 ft or more in depth shall be protected from falling by a guardrail system, fence, barricade, or cover. [1926.501(b)(7)(ii)]

g. Each employee less than 6 ft above dangerous equipment shall be protected from falling into or onto the dangerous equipment by a guardrail system or equipment guards. Also, each employee 6 ft or more above dangerous equipment shall be protected from fall hazards by a guardrail system, personal fall arrest system or safety net system. [1926.501(b)(8)(i) & (ii)]

h. Each employee working on, at, above, or near wall openings (including those with chutes attached) where the outside bottom edge of the wall opening is 6 ft or more above lower levels and the inside bottom edge of the wall opening is less than 39 in. above the walking and working surface, shall be protected from falling by the use of a guardrail system, safety net system, or personal fall arrest system. [1926.501(b)(14)]

20. Fire Protection

a. A fire protection program is to be followed throughout all phases of the construction and demolition work involved. It shall provide for effective firefighting equipment to be available without delay,

111

and designed to effectively meet all fire hazards as they occur. [1926.150(a)(1)]

b. Firefighting equipment shall be conspicuously located and readily accessible at all times, shall be periodically inspected, and be maintained in operating condition. [1926.150(a)(2) through (a)(4)]

c. Carbon tetrachloride and other toxic vaporizing liquid fire extinguishers are prohibited. [1926.150(c)(1)(vii)]

d. If the building includes the installation of automatic sprinkler protection, the installation shall closely follow the construction and be placed in service, as soon as applicable laws permit, following completion of each story. [1926.150(d)(1)]

e. A fire extinguisher, rated not less than 2A, shall be provided for each 3,000 sq ft of the protected building area, or major fraction thereof. Travel distance from any point of the protected area to the nearest fire extinguisher shall not exceed 100 ft. [1926.150(c)(1)(i)]

f. One or more fire extinguishers, rated not less than 2A, shall be provided on each floor. In multistory buildings, at least one fire extinguisher shall be located adjacent to stairway. [1926.150(c)(1)(iv)]

g. The employer shall establish an alarm system at the worksite so that employees and the local fire department can be alerted in an emergency. [1926.150(e)(1)]

21. Flag People

a. When signs, signals, and barricades do not provide necessary protection on or adjacent to a highway or street, flag person, or other appropriate traffic controls shall be provided. [1926.201(a)(1)]

b. Flag people shall be provided with and shall wear a red or orange warning garment while flagging. Warning garments worn at night shall be of reflectorized material. [1926.201(a)(4) & 1926.651(d)]

22. Flammable and Combustible Liquids

    a. No more than 25 gal shall be stored in a room outside of an approved storage cabinet. [1926.152(b)(1)]

    b. Only approved containers and portable tanks shall be used for storage and handling of flammable and combustible liquids. [1926.152(a)(1)]

    c. No more than 60 gal of flammable or 120 gal of combustible liquids shall be stored in any one storage cabinet. No more than three storage cabinets may be located in a single storage area. [1926.152(b)(1) through (b)(3)]

    d. Inside storage rooms for flammable and combustible liquids shall be of fire-resistive construction, have self-closing fire doors at all openings, 4-in. sills or depressed floors, a ventilation system that provides at least six air changes within the room per hour, and electrical wiring and equipment approved for Class I, Division 1 locations. [1926.152(b)(4)]

    e. Storage in containers outside buildings shall not exceed 1,100 gal in any one pile or area. Storage areas shall be located at least 20 ft from any building and shall be free from weeds, debris, and other combustible materials not necessary to the storage. [1926.152(c)(1) & (c)(3) through (c)(5)]. **Note:** Compliance with EPA and DOT requirements is also necessary.

    f. Flammable liquids shall be kept in closed containers when not actually in use. [1926.152(f)(1)]

    g. Conspicuous and legible signs prohibiting smoking shall be posted in service and refueling areas. [1926.152(g)(9)]

23. Floor Openings, Open Sides, and Hatchways

    a. Floor openings shall be guarded by a standard railing and toeboards or cover. In general, the railing shall be provided on all exposed sides, except at entrances to stairways. [1926.500(b)(1) & (7)]

b. Every open-sided floor or platform, 6 ft (1.8288 m) or more above adjacent floor or ground level, shall be guarded by a standard railing, or the equivalent, on all open sides except where there is entrance to a ramp, stairway, or fixed ladder. [1926.500(d)(1)]

c. Runways 4 ft (1.2192 m) or more in height shall have standard railings on all open sides, except runways 18 in. (45.72 cm) or more in width used exclusively for special purposes may have the railing on one side omitted where operating conditions necessitate. [1926.500(d)(2) & (3)]

d. Ladderway floor openings or platforms shall be guarded by standard railings with standard toeboards on all exposed sides, except at the entrance to the opening, with the passage through the railing either provided with a swinging gate or so offset that a person cannot walk directly into the opening. [1926.500(b)(2)]

e. Temporary floor openings shall have standard railings. [1926.500(b)(7)]

f. Floor holes into which persons can accidentally walk shall be guarded by either a standard railing with standard toeboard on all exposed sides, or a floor hole cover of standard strength and construction, secured to prevent accidental displacement. While the cover is not in place, the floor hole shall be protected by a standard railing. [1926.500(b)(8)]

24. Gases, Vapors, Fumes, Dusts, and Mists

a. Exposure to toxic gases, vapors, fumes, dusts, and mists at a concentration above OSHA's permissible exposure levels or above those specified in the "Threshold Limit Values of Airborne Contaminants for 1970" of the ACGIH, shall be avoided. (American Conference of Government Industrial Hygienists, 1330 Kemper Meadow Drive, Cincinnati, OH 45240. Telephone #(513) 742-2020). [1926.55(a)]

b. Administrative or engineering controls must be implemented whenever feasible to comply with threshold limit values (TLVs). 1926.55(b)]

c. When engineering and administrative controls are not feasible to achieve full compliance, protective equipment or other protective measures shall be used to keep the exposure of employees to air contaminants within the limits prescribed. Any equipment and technical measures used for this purpose must first be approved for each particular use by a competent industrial hygienist or other technically qualified person. [1926.55(b)]

d. Whenever respirators are used, the employer shall comply with the respirator requirements. [1926.103]

25. General Duty Clause

Hazardous conditions or practices not covered in an OSHA standard may be covered under Section 5(a)(1) of the Occupational Safety and Health Act of 1970, which states: "Each employer shall furnish to each of his employees employment and a place of employment [that] are free from recognized hazards that are causing or are likely to cause death or serious physical harm to his employees."

26. General Safety and Health Requirements

a. The employer shall initiate and maintain such programs as may be necessary to provide for frequent and regular inspections of the job site, materials, and equipment by a competent person. A safety program is required. [1926.20(b)]

b. Training. The employer shall instruct each employee in the recognition and avoidance of unsafe conditions and in the regulations applicable to his work environment to control or eliminate any hazards or other exposure to illness or injury. [1926.21(b)(2)]

c. The use of any machinery, tool, material, or equipment that is not in compliance with any applicable requirement of Part 1926 is prohibited. [1926.20(b)(3)]

    d. The employer shall permit only those employees qualified by training or experience to operate equipment and machinery. [1926.20(b)(4)]

27. Hand Tools

    a. Employers shall not issue or permit the use of unsafe hand tools. [1926.301(a)]

    b. Wrenches shall not be used when jaws are sprung to the point that slippage occurs. Impact tools shall be kept free of mushroomed heads. The wooden handles of tools shall be kept free of splinters or cracks and shall be kept tight in the tool. [1926.301(b), (c), & (d)]

    c. Electric power-operated tools shall either be approved double-insulated, or be properly grounded, and used with ground fault circuit interrupters. [1926.302(a) & 1926.404(b)(1)(ii)]

28. Hazard Communication

    a. The purpose of this standard is to ensure that the hazards of all chemicals produced or imported are evaluated, and that information concerning their hazards is transmitted to employers and employees. This transmittal of information is to be accomplished by means of comprehensive hazard communication programs, which are to include container labeling and other forms of warning, material safety data sheets and employee training. [1926.59(a)(1)]

    b. Employers shall develop, implement and maintain at the work place, a written hazard communication program for their work places. Employers must inform their employees of the availability of the program, including the required lists of hazardous chemicals, and material safety data sheets required. [1926.59(e)(1)]

    c. The employer shall ensure that each container of hazardous chemicals in the workplace is labeled, tagged or marked with the identity of the hazardous chemicals contained therein; and must show hazard warnings appropriate for employee protection. [1926.59(f)(5)]

d. Chemical manufacturers and importers shall obtain or develop a material safety data sheet for each hazardous chemical they produce or import. Employers shall have a material safety data sheet for each hazardous chemical which they use and MSDSs must be available at the jobsite. [1926.59(g)(1)]

e. Employers shall provide employees with information and training on hazardous chemicals in their work area at the time of their initial assignment, and whenever a new hazard is introduced into their work area. Employers shall also provide employees with information on any operations in their work area where hazardous chemicals are present; and the location and availability of the written hazard communication program, including the required list(s) of hazardous chemicals, and material safety data sheets required by the standard. [1926.59(h)(1)]

f. Employers who produce, use, or store hazardous chemicals at multiemployer workplaces shall additionally ensure that their hazard communication program includes the methods the employer will use to provide other employers with a copy of the material safety data sheet for hazardous chemicals other employers employees may be exposed to while working; the methods the employer will use to inform other employers of any precautionary measures for the protection of employees; and the methods the employer will use to inform the other employers of the labeling system used in the workplace. [1926.59(e)(2)]

29. Head Protection

a. Head protective equipment (hard hats) shall be worn in areas where there is a possible danger of head injuries from impact, flying or falling objects, or electrical shock and burns. [1926.100(a)]

b. Hard hats for protection against impact and penetration of falling and flying objects shall meet the requirements of ANSI Z89.1-1969. [1926.100(b)]

c. Helmets for protection against electrical shock and burns shall meet the requirements of ANSI Z89.2-1971. [1926.100(c)]

30. Hearing Protection

   a. Feasible engineering or administrative controls shall be utilized to protect employees against sound levels in excess of those shown in Table 16.1 (OSHA Table D-2) [1926.52(b)]

   b. When engineering or administrative controls fail to reduce sound levels within the limits of Table 16.1 (OSHA Table D-2), ear protective devices shall be provided and used. [1926.52(b) & 1926.101(a)]

   c. Exposure to impulsive or impact noise should not exceed 140-dB peak sound pressure level. [1926.52(e)]

   d. In all cases where the sound levels exceed the values shown in Table 16.1 (OSHA Table D-2) of the Safety and Health Standards, a continuing, effective hearing conservation program shall be administered. [1926.52(d)(1)]

   e. Table D-2 Permissible Noise Exposures) [1926.52(d)(1)]

Table 16.1 (OSHA Table D–2)

Permissible Noise Exposures	
Duration Per Day in Hours	**Sound Level dBA Slow Response**
8	90
6	92
4	95
3	97
2	100
1½	102
1	105
½	110
¼ or less	115

f. Plain cotton is not an acceptable protective device. [1926.101(c)]

31. Heating Devices, Temporary

   a. Fresh air shall be supplied in sufficient quantities to maintain the health and safety of workers. [1926.154(a)(1)]

   b. Solid fuel salamanders are prohibited in buildings and on scaffolds. [1926.154(d)]

32. Hoists, Material and Personnel

   a. The employer shall comply with the manufacturer's specifications and limitations. [1926.552(a)(1)]

   b. Rated load capacities, recommended operating speeds, and special hazard warnings or instructions shall be posted on cars and platforms. [1926.552(a)(2)]

   c. Hoistway entrances of material hoists shall be protected by substantial full width gates or bars. [1926.552(b)(2)]

   d. Hoistway doors or gates of personnel hoist shall be not less than 6 ft, 6 in. high, and be protected with mechanical locks which cannot be operated from the landing side and are accessible only to persons on the car. [1926.552(c)(4)]

   e. Overhead protective coverings shall be provided on the top of the hoist cage or platform. [1926.552(b)(3) & (c)(7)]

   f. All material hoists shall conform to the requirements of ANSI A10.5-1969, Safety Requirements for Material Hoists. [1926.552(b)(8)]

33. Hooks (See Wire Ropes, Chains, Hooks, and so on)

34. Housekeeping

   a. Form and scrap lumber with protruding nails and all other debris, shall be kept clear from all work areas. [1926.25(a)]

   b. Combustible scrap and debris shall be removed at regular intervals. [1926.25(b)]

c. Containers shall be provided for collection and separation of all refuse. Covers shall be provided on containers used for flammable or harmful substances. [1926.25(c)]

d. Wastes shall be disposed of at frequent intervals. [1926.25(c)]

35. Illumination

a. Construction areas, ramps, runways, corridors, offices, shops, and storage areas shall be lighted to not less than the minimum illumination intensities listed in Table 16.2 (OSHA Table D-3) while any work is in progress.

b. Table D-3 Minimum Illumination Intensities in Foot-Candles

Table 16.2 (OSHA Table D–3)

Minimum Illumination Intensities in Foot-Candles	
Foot-Candles	**Area of Operation**
5	General construction area lighting.
3	General construction areas, concrete placement, excavation, waste areas, accessways, active storage areas, loading platforms, refueling, and field maintenance areas.
5	Indoor: warehouses, corridors, hallways, and exitways.
5	Tunnels, shafts, and general underground work areas. (Exceptions: minimum of 10 foot-candles is required at tunnel and shaft heading during drilling, mucking, and scaling. Bureau of Mines approved cap lights shall be acceptable for use in the tunnel heading.)
10	General construction plant and shops (e.g., batch plants, screening plants, mechanical and electrical equipment rooms, carpenters shops, rigging lofts and active storerooms, barracks or living quarters, locker or dressing rooms, mess halls, indoor toilets, and workrooms).
10	First-aid stations, infirmaries, and offices.

36. Jointers

    a. Each hand-fed planer and jointer with a horizontal head shall be equipped with a cylindrical cutting head. The opening in the table shall be kept as small as possible. [1926.304(f)]

    b. Each hand-fed jointer with a horizontal cutting head shall have an automatic guard which will cover the section of the head on the working side of the fence or cage. [1926.304(f)]

    c. A jointer guard shall automatically adjust itself to cover the unused portion of the head, and shall remain in contact with the material at all times. [1926.304(f)]

    d. Each hand-fed jointer with horizontal cutting head shall have a guard which will cover the section of the head back of the cage or fence. [1926.304(f)]

37. Ladders

    a. Portable and fixed ladders with structural defects such as broken or missing rungs, cleats or steps, broken or split rails, or corroded components shall be withdrawn from service by immediately tagging "Do Not Use" or marking in a manner that identifies them as defective; or blocked (such as with a plywood attachment that spans several rungs). Repairs must restore ladder to its original design criteria. [1926.1053(b)(16) through (17)]

    b. Portable ladders shall be placed on a substantial base, have clear access at top and bottom and be placed at an angle so the horizontal distance from the wall or top support to the foot of the ladder is approximately one-quarter the working length of the ladder. Portable ladders used for access to an upper landing surface must extend a minimum of 3 ft (0.9 m) above the landing surface, or where not practical, be provided with grab rails and be secured against movement while in use. [1926.1053(b)(1) & (5)(i)]

c. Ladders must have nonconductive siderails if they are used where the worker or the ladder could contact energized electrical conductors or equipment. [1926.1053(b)(12)]

d. Job-made ladders shall be constructed for their intended use. Cleats shall be uniformly spaced not less than 10 in. (0.25 cm) apart, nor more than 14 in. (36 cm) apart, along the handrails. [1926.1053(a)(3)(i)]

e. A ladder (or stairway) must be provided at all work points of access where there is a break in elevation of 19 in. (48 cm) or more except if a suitable ramp, runway, embankment, or personnel hoist is provided to give safe access to all elevations. [1926.1051(a)]

f. Non–self-supporting ladders must be used at an angle where the horizontal distance from the top support to the foot of the ladder is approximately one-quarter of the working length of the ladder. Wood job-made ladders with spliced side rails must be used at an angle where the horizontal distance is one-eighth the working length of the ladder.

   • Fixed ladders must be used at a pitch no greater than 90 degrees from the horizontal, measured from the back side of the ladder.

   • Ladders must be used only on stable and level surfaces unless secured to prevent accidental movement.

   • Ladders must not be used on slippery surfaces unless secured or provided with slip-resistant feet to prevent accidental movement. Slip-resistant feet must not be used as a substitute for the care in placing, lashing, or holding a ladder upon a slippery surface. [1926.1053(b)(5) through (b)(7)]

g. Employers must provide a training program for each employee using ladders and stairways. The program must enable each employee to recognize hazards related to ladders and stairways and to use proper procedures to minimize these hazards. For example, employers must ensure that each employee is trained by a competent person in the following areas, as applicable:

- The nature of fall hazards in the work area;
- The correct procedures for erecting, maintaining, and disassembling the fall protection systems to be used;
- The proper construction, use, placement, and care in handling of all stairways and ladders; and
- The maximum intended load-carrying capacities of ladders used.

In addition, retraining must be provided for each employee as necessary so that the employee maintains the understanding and knowledge acquired through compliance with the standard. [1926.1060(a) & (b)]

38. Lasers

a. Only qualified and trained employees shall be assigned to install, adjust, and operate laser equipment. [1926.54(a)]

b. Employees shall wear proper eye protection where there is a potential exposure to laser light greater than 0.005 watt (5 milliwatts/sq cm). [1926.54(c)]

c. Beam shutters or caps shall be utilized, or the laser turned off, when laser transmission is not actually required. When the laser is left unattended for a substantial period of time, such as during lunch hour, overnight, or at change of shifts, the laser shall be turned off. [1926.54(e)]

d. Employees shall not be exposed to light intensities above: direct staring—1 micro-watt per square centimeter; incidental observing—1 milliwatt per square centimeter; diffused reflected light—2 1/2 watts per square centimeter. Employees shall not be exposed to microwave power densities in excess of

10 milliwatts per square centimeter. [1926.54(j)(1) through (j)(3) & (l)]

    e. Employees shall not be exposed to microwave power densities in excess of 10 milliwatts per square centimeter. [1926.54(l)]

39. Liquified Petroleum Gas (LPG)

    a. Each system shall have containers, valves, connectors, manifold valve assemblies, and regulators of an approved type. [1926.153(a)(1)]

    b. All cylinders shall meet DOT specifications. [1926.153(a)(2)]

    c. Every container and vaporizer shall be provided with one or more approved safety relief valves or devices. [1926.153(d)(1)]

    d. Containers shall be placed upright on firm foundations or otherwise firmly secured. [1926.153(g) & (h)(11)]

    e. Portable heaters shall be equipped with an approved automatic device to shut off the flow of gas in the event of flame failure. [1926.153(h)(8)]

    f. Cylinder shall be equipped with an excess flow valve to minimize the flow of gas in the event the fuel line becomes ruptured. [1926.153(j)]

    g. Storage of LPG within buildings is prohibited. [1926.153(l)]

    h. Storage locations shall have at least one approved portable fire extinguisher, rated not less than 20-B:C. [1926.153(i)(2)]

40. Medical Services and First-Aid

    a. The employer shall ensure the availability of medical personnel for advice and consultation on matters of occupational health. [1926.50(a)]

    b. When a medical facility is not reasonably accessible for the treatment of injured employees, a person trained to render first aid shall be available at the work site. [1926.50(c)]

c. First-aid supplies approved by the consulting physician shall be readily available. [1926.50(d)(1)]

d. The telephone numbers of the physicians, hospitals, or ambulances shall be conspicuously posted. [1926.50(f)]

41. Motor Vehicles and Mechanized Equipment

a. All vehicles in use shall be checked at the beginning of each shift to ensure that all parts, equipment, and accessories that affect safe operation are in proper operating condition and free from defects. All defects shall be corrected before the vehicle is placed in service. [1926.601(b)(14)]

b. No employer shall use any motor vehicle, earthmoving, or compacting equipment having an obstructed view to the rear unless:

- The vehicle has a reverse signal alarm distinguishable from the surrounding noise level; or

- The vehicle is backed up only when an observer signals that it is safe to do so. [1926.601(b)(4) & 1926.602(a)(9)]

c. Heavy machinery, equipment, or parts thereof which are suspended or held aloft shall be substantially blocked to prevent falling or shifting before employees are permitted to work under or between them. [1926.600(a)(3)(i)]

42. Noise (See Hearing Protection)

43. Personal Protective Equipment

a. The employer is responsible for requiring the wearing of appropriate personal protective equipment in all operations where there is an exposure to hazardous conditions or where the need is indicated for using such equipment to reduce the hazard to the employees. [1926.28(a)]

b. Lifelines, safety belts, and lanyards shall be used only for employee safeguarding. [1926.104(a)]

c. Employees working over or near water, where the danger of drowning exists, shall be provided with U.S. Coast Guard–approved life jackets or buoyant work vests. [1926.106(a)]

44. Power-Actuated Tools

a. Only trained employees shall be allowed to operate power-actuated tools. [1926.302(e)(1)]

b. All power-actuated tools shall be tested daily before use and all defects discovered before or during use shall be corrected. [1926.302(e)(2) & (e)(3)]

c. Tools shall not be loaded until immediately before use. Loaded tools shall not be left unattended. [1926.302(e)(5) & (e)(6)]

45. Power Transmission and Distribution

a. Existing conditions shall be determined before starting work, by an inspection or a test. Such conditions shall include, but not be limited to, energized lines and equipment, condition of poles, and the location of circuits and equipment including power and communications, CATV, and fire alarm circuits. [1926.950(b)(1)]

b. Electric equipment and lines shall be considered energized until determined otherwise by testing or until grounding. [1926.950(b)(2) & 1926.954(a)]

c. Operating voltage of equipment and lines shall be determined before working on or near energized parts. [1926.950(b)(3)]

d. Rubber protective equipment shall comply with the provisions of the ANSI J6 series and shall be visually inspected before use. [1926.951(a)(1)]

46. Power Transmission, Mechanical

a. Belts, gears, shafts, pulleys, sprockets, spindles, drums, flywheels, chains, or other reciprocating, rotating, or moving parts of equipment shall be guarded if such parts are exposed to contact by employees or otherwise constitute a hazard. [1926.300(b)(2)]

b. Guarding shall meet the requirement of ANSI B15.1-1953 (R 1958), "Safety Code for Mechanical Power Transmission Apparatus." [1926.300(b)(2)]

47. Process Safety Management of Highly Hazardous Chemicals

   a. Employers shall develop a written plan of action regarding employee participation and consult with employees and their representatives on the conduct and development of process hazards analyses and on the development of the other elements of process safety management. [1926.64(c)(1) & (2)]

   The employer shall complete a compilation of written process safety information prior to conducting a process hazard analysis (PHA). [1926.64(d)]

   b. The employer shall perform a process hazard analysis appropriate to the complexity of the company's processes and shall identify, evaluate, and control the hazards involved in the process. [1926.64(e)(1)]

   c. The employer shall develop and implement written operating procedures that provide clear instructions for safely conducting activities involved in each covered process consistent with process safety infor-mation. [1926.64(f)(1)]

   d. Each employee presently involved in operating a process, and each employee before being involved in operating a newly assigned process, shall be trained in an overview of the process and in the operating procedures specified in paragraph (f). [1926.64(g)(1)]

   e. The employer, when selecting a contractor, shall obtain and evaluate information regarding the contract employer's safety performance and programs. [1926.64(h)(2)(i)]

   f. The contract employer shall assure that each contract employee is trained in the work practices necessary to safely perform his/her job. [1926.64(h)(3)(i)]

g. The employer shall perform a pre-startup safety review for new facilities and for modified facilities when the modification is significant enough to require a change in the process safety information. [1926.64(i)(1)]

h. The employer shall establish and implement written procedures to maintain the on going integrity of process equipment. [1926.64(i)(2)]

i. The employer shall establish and implement written procedures to manage changes to process chemicals, technology, equipment, and procedures, and changes to facilities that affect a covered process. [1926.64(i)(1)]

48. Radiation, Ionizing

a. Pertinent provisions of the Atomic Energy Commission's Standards for Protection Against Radiation (10 CFR part 20), relating to protection against occupational radiation exposure, shall apply. [1926.53(a)]

b. Any activity that involves the use of radioactive materials or X rays, whether or not under license from the Atomic Energy Commission, shall be performed by competent persons specially trained in the proper and safe operation of such equipment. [1926.53(b)]

49. Railings

a. A standard railing shall consist of top rail, intermediate rail, toeboard, and posts. It shall have a vertical height of approximately 42 in. ±3 inches from upper surface of top rail to the floor, platform, and so on. [1926.500(f)(1)]

b. The top rail of a railing shall be smooth surfaced, with a strength to withstand at least 200 lb, the minimum requirement applied in an outward or downward direction at any point on the top rail, with a minimum of deflection. The intermediate rail shall be approximately halfway between the top rail and floor. [1926.500(f)(1)]

c. A stair railing shall be of construction similar to a standard railing, but the vertical height shall be not less than 30 in. from the upper surface of the top rail to surface of tread in line with face of riser at forward edge of tread. [1926.500(f)(2)]

50. Respiratory Protection

a. In emergencies or when feasible engineering or administrative controls are not effective in controlling toxic substances, appropriate respiratory protective equipment shall be provided by the employer and shall be used. [1926.103(a)(1)]

b. Respiratory protective devices shall be approved by the Mine Safety and Health Administration/National Institute for Occupational Safety and Health or acceptable to the U.S. Department of Labor for the specific containment to which the employee is exposed. [1926.103(a)(2)]

c. Respiratory protective devices shall be appropriate for the hazardous material involved and the extent and nature of the work requirements and conditions. [1926.103(b)(1) & (b)(2)]

d. Employees required to use respiratory protective devices shall be thoroughly trained in their use. [1926.103(c)(1)]

e. Respiratory protective equipment shall be inspected regularly and maintained in good condition. [1926.103(c)(2)]

f. Written standard operating procedures for the selection and use of respirators should be established. [1910.134(b)]

h. Persons should not be assigned to tasks requiring use of respirators, unless it has been determined by a physician that they are physically able to use a respirator. [1910.134(a)(10)]

51. Rollover Protective Structures (ROPS)

Rollover protective structures (ROPS) applies to the following types of materials handling equipment: to all rubber-tired dozers, wheel-type agricultural and industrial tractors, crawler tractors, crawler-type loaders, and motor graders, with or without attachments, that are used in construction work. This requirement does not apply to sideboom pipelaying tractors. [1926.1000(a)(1)]

52. Safety Nets

    a. Safety nets shall be provided when workplaces are more than 25 ft above the surface where the use of ladders, scaffolds, catch platforms, temporary floors, safety lines, or safety belts is impractical. [1926.105(a)]

    b. Where nets are required, operations shall not be undertaken until the net is in place and has been tested. [1926.105(b)]

53. Saws, Band

    a. All portions of band saw blades shall be enclosed or guarded, except for the working portion of the blade between the bottom of the guide rolls and the table. [1926.304(f)]

    b. Band saw wheels shall be fully encased. [1926.304(f)]

54. Saws, Portable Circular

Portable, power-driven circular saws shall be equipped with guards above and below the base plate or shoe. The lower guard shall cover the saw to the depth of the teeth, except for the minimum arc required to allow proper retraction and contact with the work, and shall automatically return to the covering position when the blade is removed from the work. [1926.304(d)]

55. Saws, Radial

    a. Radial saws shall have an upper guard, which completely encloses the upper half of the saw blade. The sides of the lower exposed portion of the blade shall be guarded by a device that will automatically adjust to the thickness of and remain in contact with the material being cut. [1926.304(g)(1)]

b. Radial saws used for ripping shall have nonkickback fingers or dogs. [1926.304(f)]

c. Radial saws shall be installed so that the cutting head will return to the starting position when released by the operator. [1926.304(f)]

56. Saws, Swing or Sliding Cut-Off

a. All swing or sliding cut-off saws shall be provided with a hood that will completely enclose the upper half of the saw blade. [1926.304(d) & (f)]

b. Limit stops shall be provided to prevent swing or sliding type cut-off saws from extending beyond the front or back edges of the table. [1926.304(f)]

c. Each swing or sliding cut-off saw shall be provided with an effective device to return the saw automatically to the back of the table when released at any point of its travel. [1926.304(f)]

d. Inverted sawing of sliding cut-off saws shall be provided with a hood that will cover the part of the saw that protrudes above the top of the table or material being cut. [1926.304(f)]

57. Saws, Table

a. Circular table saws shall have a hood over the portion of the saw above the table, so mounted that the hood will automatically adjust itself to the thickness of and remain in contact with the material being cut. [1926.304(c)(1),(d) & (h)(1)]

b. Circular table saws shall have a spreader aligned with the blade, spaced no more than 1/2 in. behind the largest blade mounted in the saw. This provision does not apply when grooving, dadoing, or rabbiting. [1926.304(f)]

c. Circular table saws used for ripping shall have nonkickback fingers or dogs. [1926.304(f)]

d. Feeder attachments shall have the feed rolls or other moving parts covered or guarded so as to protect the operator from hazardous points. [1926.304(c)]

58. Scaffolds (General)

   a. Scaffolds and scaffold components shall be capable of supporting its own weight and at least 4 times the maximum intended load applied or transmitted to it. [1926.451(a)(1)]

   b. Suspension ropes and hardware used on suspension scaffolds shall be capable of supporting 6 times the maximum intended load applied or transmitted to the rope. [1926.451(a)(3) & (4)]

   c. Scaffolds shall be designed by a qualified person and shall be constructed and loaded in accordance with that design. [1926.451(a)(6)]

   d. Scaffolds shall be braced to prevent tipping and shall be erected on sound, rigid footing, capable of carrying the maximum intended load without settling or displacement. [1926.451(c)(2)(i) & (ii)]

   e. Each employee on a scaffold more than 10 feet above a lower level shall be protected from falling to lower levels by guardrails and/or personal fall arrest systems. [1926.451(g)]

   f. Employees are required to wear hardhats. Where there is a danger of tools, materials, or other equipment falling from a scaffold and striking employees below, the following provisions apply:

      • area below shall be barricaded and employees shall not be permitted in the hazard area; or
      • toeboards or screening shall be erected; or
      • canopy structure, debris net, or catch platform shall be erected. [1926.451(h)(1) & (2)

   g. Scaffolds and scaffold components shall not be loaded in excess to their maximum intended loads or rated capacities, whichever is less. [1926.451(f)(1)]

   h. Scaffolds and scaffold components shall be inspected for visible defects by a competent person before each work shift, and after any occurrence which could affect a scaffold's integrity. [1926.451(f)(3)]

i. Any scaffold or scaffold component that is damaged shall be replaced, repaired, braced, or removed from service. {1926.451(f)(4)]

j. Scaffold planking shall be overlapped a minimum of 12 inches or secured from movement, and planks shall extend over their end supports not less than 6 inches or more than 12 inches. [1926.451(b)(4), (5), & (7)]

k. Scaffold platforms more than 2 feet above or below the point of access must be equipped with a ladder, stairway, ramp, or other safe means of access. [1926.451(e)]

l. Scaffolds shall not be moved horizontally while employees are on them, unless specifically designed for such movement. [1926.451(f)(5)]

m. Scaffolds shall not be erected, used, dismantled, altered, or moved such that they or any conductive material handled on them might come in contact with energized power lines. [1926.451(f)(6)]

n. Scaffolds shall be erected, moved, dismantled, or altered only under the supervision and direction of a qualified competent person. Such activities shall be performed only by experienced and trained employees. [1926.451(f)(7)]

o. Ladders, boxes, barrels, or other makeshift devices shall not be used on scaffolds to increase the working level of employees, except on large area scaffolds. [1926.451(c)(iv) & (v)]

p. Front-end loaders, forklifts, and other equipment shall not be used to support scaffold platforms unless they have been specifically designed by the manufacturer for such use. [1926.451(c)(iv) & (v)]

59. Scaffolds (supported or mobile)

a. Supported scaffolds with a height to base width (including outriggers, if used) ratio of 4:1 shall be restrained from tipping. Restraints shall be installed

133

according to manufacturer's recommendation or at the closest horizontal member to the 4:1 height. [1926.451(c)(1)]

b. Restraints shall be installed every 20 feet or less vertically for scaffolds 3 feet wide or less, and every 26 feet or less for scaffolds greater than 3 feet wide. [1926.451(c)(ii)]

c. Restraints shall be installed at each end of the scaffold and at horizontal intervals not to exceed 30 feet. [1926.451(c)(ii)]

d. Scaffolds shall be plumb, level and square. All brace connections shall be secured. [1926.452(c)(2) & (w)(1)]

e. Mobile scaffold casters and wheels shall be locked with positive wheel locks or equivalent means, to prevent movement of the scaffold while in use. [1926.452(w)(2)]

f. Manual force used to move mobile scaffolds shall be applied as close to the base as practicable, but not more than 5 feet above the supporting surface. [1926.452(w)(3)]

g. Equipment shall not be used to move mobile scaffolds unless the scaffold is designed to be propelled by equipment. [1926.452(w)(4)]

h. Employees shall not ride on mobile scaffolds unless:
   • on a smooth surface of less than 3 degrees;
   • the height to base ratio is less than 2:1;
   • outriggers are installed when used;
   • the system is designed for riders and propelled by a power system;
   • no employee is on any part of the scaffold which extends outward beyond the wheels, casters, or other supports. [1926.452(w)(6)]

i. Caster stems and wheels for mobile scaffolds shall be secured to the frame of the scaffold. [1926.452(w)(9)]

j. Employees working on mobile scaffolds shall be made aware of the move. [1926.452(w)(10)]

60. Scaffolds (suspended)

   a. All suspended scaffold support devices shall rest on surfaces capable of supporting at least 4 times the load imposed on them. [1926.451(d)(1)]

   b. The scaffold shall be evaluated by a competent person before the scaffold is used, prior to each work-shift and after every occurrence which could affect the scaffold's integrity. [1926.451(d)(3)(i) & (d)(10)]

   c. Only counterweights specifically designed as counterweights should be used. Sand, gravel, flowable materials, concrete blocks, bricks, roofing materials, and other similar materials shall not be used as counterweights. [1926.451(d)(ii) & (iii)]

   d. Work on scaffolds is prohibited during storms or high winds unless a competent person has determined that it is safe and those employees are protected by a personal fall arrest system or wind screens. [1926.451(f)(12)]

   e. Employees on single-point, two-point, boatswain's chair, catenary, float, needle-beam, ladder jack, scaffolds and chicken ladders shall be protected by personal fall arrest systems in addition to any other fall protection required. [1926.451(g)(i), (ii), & (iii)]

61. Scaffolds (training)

   a. By virtue of training or experience the competent person must be capable of identifying existing or predictable hazards in reference to erecting, inspecting, using, and dismantling scaffolds. [1926.450(b)]

   b. Each employee who performs work while on a scaffold must be trained to recognize and control scaffold hazards by a qualified person knowledgeable about the hazards associated with the type of scaffold being used. The training shall include:

135

- the nature of electrical, fall, and falling object hazards;
- procedures for installing, using, and disassembling the fall protecction system and falling object protection system;
- the proper use of scaffolds and handling materials on the scaffold;
- the maximum intended load and load carrying capacities of the scaffold;
- other pertinent information that will provide the worker with a safe place to work. [1926.454(a)]

c. Employees involved in erecting, dismantling, moving, or inspecting scaffolds must be trained by a competent person to recognize hazards. The training shall include:
- the nature of the scaffold hazards;
- the correct procedures for erecting, disassembling, moving, repairing, inspecting, and maintaining the scaffold;
- design criteria and other pertinent information. [1926.454(b)]

d. Retraining is required when changes at the worksite or to the scaffold present a hazard, when changes to worker protection are made, or when employee(s) has not retained the training that was provided. [1926.454(c)]

62. Stairs
a. A stairway or ladder must be provided at all worker points of access where there is a break in elevation of 19 in. (48 cm) or more and no ramp, runway, embankment, or personnel hoist is provided. [1926.1051(a)]

b. Except during construction of the actual stairway, skeleton metal frame structures and steps must not be used (where treads and/or landings are to be installed at a later date), unless the stairs are fitted with secured temporary treads and landings. [1926.1052(b)(2)]

c. When there is only one point of access between levels, it must be kept clear to permit free passage by workers. If free passage becomes restricted, a second point of access must be provided and used. [1926.1051(a)(3)]

d. When there are more than two points of access between levels, at least one point of access must be kept clear. [1926.1051(a)(4)]

e. All stairway and ladder fall protection systems required by these rules must be installed, and all duties required by the stairway and ladder rules must be performed before employees begin work that requires them to use stairways or ladders and their respective fall protection systems. [1926.1051(b)]

f. Stairways that will not be a permanent part of the structure on which construction work is performed must have landings at least 30 in. deep and 22 in. wide (76 × 56 cm) at every 12 ft (2.7 m) or less of vertical rise. [1926.1052(a)(1)]

g. Stairways must be installed at an angle of at least 30 degrees, and no more than 50 degrees, from the horizontal. [1926.1052(a)(2)]

h. Variations in riser height, or stair tread, or landing depth, must not exceed $1/4$ in. in any stairway system, including any foundation structure used as one or more treads of the stairs. [1926.1052(a)(3)]

i. Where doors or gates open directly onto a stairway, a platform must be provided that extends at least 20 in. (51 cm) beyond the swing of the door. [1926.1052(a)(4)]

j. Except during construction of the actual stairway, stairways with metal pan landings and treads must not be used where the treads and/or landings have not been filled in with concrete or other material, unless the pans of the stairs and/or landings are temporarily filled in with wood or other material. All treads and landings must be replaced when worn below the top edge of the pan. [1926.1052(b)(1)]

k. Stairways having four or more risers, or rising more than 30 in. (76 cm) in height, whichever is less, must have at least one handrail. A stairrail also must be installed along each unprotected side or edge. When the top edge of a stairrail system also serves as a handrail, the height of the top edge must not be more than 37 in. (94 cm) nor less than 36 in. (91.5 cm) from the upper surface of the stairrail to the surface of the tread. [1926.1052(c)(1)]

l. Winding or spiral stairways must be equipped with a handrail to prevent using areas where the tread width is less than 6 in. (15 cm). [1926.1052(c)(2)]

m. Stairrails installed after March 15, 1991 must not be less than 36 in. (91.5 cm). [1926.1052(c)(3)]

n. Midrails, screens, mesh, intermediate vertical members, or equivalent intermediate structural members must be provided between the top rail and stairway steps of the stairrail system. [1926.1052(c)(4)]

o. Midrails, when used, must be located midway between the top of the stairrail system and the stairway steps. [1926.1052(c)(4)(i)]

p. The height of handrails must be not more than 37 in. (94 cm) nor less than 30 in. (76 cm) from the upper surface of the handrail to the surface of the tread. [1926.1052(c)(6)]

q. The height of the top edge of a stairrail system used as a handrail must not be more than 37 in. (94 cm) nor less than 36 in. (91.5 cm) from the upper surface of the stairrail system to the surface of the tread. [1926.1052(c)(7)]

r. Temporary handrails must have a minimum clearance of 3 in. (8 cm) between the handrail and walls, stair-rails systems, and other objects. [1926.1052(c)(11)]

s. Unprotected sides and edges of stairway landings must be provided with standard 42 in. (1.1 m) guardrail systems. [1926.1052(c)(12)]

63. Steel Erection

a. Permanent floors shall be installed so there is not more than eight stories between the erection floor and the uppermost permanent floor, except when structural integrity is maintained by the design. [1926.750(a)(1)]

b. During skeleton steel erection, a tightly planked temporary floor shall be maintained within two stories or 30 ft, whichever is less below and directly under the portion of each tier of beams on which any work is being performed. [1926.750(b)(2)(i)]

c. During skeleton steel erection, where the requirements of the preceding paragraph cannot be met, and where scaffolds are not used, safety nets shall be installed and maintained whenever the potential fall distance exceeds two stories or 25 ft. [1926.750(b)(1)(ii)]

d. A safety railing of ½-in. wire rope or equivalent shall be installed around the perimeter of all temporarily floored buildings, approximately 42 in. high, during structural steel assembly. [1926.750(b)(1)(iii)]

e. When placing structural members, the load shall not be released from the hoisting line until the member is secured by at least two bolts, or the equivalent, at each connection, drawn up wrench tight. [1926.751(a)]

64. Storage

a. All materials stored in tiers shall be secured to prevent sliding, falling, or collapse. [1926.250(a)(1)]

b. Aisles and passageways shall be kept clear and in good repair. [1926.250(a)(3)]

c. Storage of materials shall not obstruct exits. [1926.151(d)(1)]

d. Materials shall be stored with due regard to their fire characteristics. [1926.151(d)(2)]

e. Weeds and grass in outside storage areas shall be kept under control. [1926.151(c)(3)]

65. Tire Cages

A safety tire rack, cage, or equivalent protection shall be provided and used when inflating, mounting, or dismounting tires installed on split rims, or rims equipped with locking rings or similar devices. [1926.600(a)(2)]

66. Toilets

a. Toilets shall be provided according to the following: 20 or fewer persons—one facility; 20 or more persons—one toilet seat and one urinal per 40 persons; 200 or more persons—one toilet seat and one urinal per 50 workers. [1926.51(c)(1)]

b. This requirement does not apply to mobile crews having transportation readily available to nearby toilet facilities. [1926.51(c)(4)]

67. Underground Construction

a. The employer shall provide and maintain safe means of access and egress to all work stations. [1926.800(b)]

b. The employer shall control access to all openings to prevent unauthorized entry underground. Unused chutes, manways, or other openings shall be tightly covered, bulkheaded, or fenced off, and shall be posted with signs indicating "Keep Out" or similar language. Completed or unused sections of the underground facility shall be barricaded. [1926.800(b)(3)]

c. Unless underground facilities are sufficiently completed, the employer shall maintain a check-in/check-out procedure that will ensure that above-ground designated personnel can determine an accurate count of the number of persons underground in the event of an emergency. [1926.800(c)]

d. All employees shall be instructed to recognize and avoid hazards associated with underground construction activities. [1926.800(d)]

e. Hazardous classifications are potentially gassy and gassy operations. [1926.800(h)]

f. The employer shall assign a competent person to perform all air monitoring to determine proper ventilation and quantitative measurements of potentially hazardous gases. [1926.800(j)(1)(i)(A)]

g. Fresh air shall be supplied to all underground work areas in sufficient quantities to prevent dangerous or harmful accumulation of dust, fumes, mists, vapors, or gases. [1926.800(k)]

68. Wall Openings (See Fall Protection)

69. Washing Facilities

a. The employer shall provide adequate washing facilities for employees engaged in operations involving harmful substances. [1926.51(f)]

b. Washing facilities shall be in near proximity to the worksite and shall be so equipped as to enable employees to remove all harmful substances. [1926.51(f)]

70. Welding, Cutting, and Heating

a. Employers shall instruct employees in the safe use of welding equipment. [1926.350(d) & 1926.351(d)]

b. Proper precautions (isolating welding and cutting, removing fire hazards from the vicinity, providing a fire watch, and so on) for fire prevention shall be taken in areas where welding, cutting or heating shall be done where the application of flammable paints, or the presence of other flammable compounds, or heavy dust concentrations creates a fire hazard. [1926.532(a), (b), (c), & (f)]

c. All welding and cutting operations shall be shielded by noncombustible or flameproof shields to protect employees from direct arc rays. [1926.351(e)]

d. When electrode holders are to be left unattended, the electrodes shall be removed and the holder shall be placed or protected so that they cannot make electrical contact with employees or conducting objects. [1926.351(d)(1)]

e. All arc welding and cutting cables shall be completely insulated and be capable of handling the maximum current requirements for the job. There shall be no repairs or splices within ten feet of the electrode holder, except where splices are insulated equal to the insulation of the cable. Defective cable shall be repaired or replaced. [1926.351(b)(1), (b)(2) & (b)(4)]

f. Fuel gas and oxygen hose shall be easily distinguishable and shall not be interchangeable. Hoses shall be inspected at the beginning of each shift and shall be repaired or replaced if defective. [1926.350(f)(1) & (f)(3)]

g. General mechanical or local exhaust ventilation or air line respirators shall be provided, as required, when welding, cutting or heating:

- zinc-, lead-, cadmium-, mercury-, or beryllium-bearing, based or coated materials in enclosed spaces;
- stainless steel with inert-gas equipment;
- in confined spaces; or
- where an unusual condition can cause an unsafe accumulation of contaminants. [1926.353(b)(1), (c)(1) through (c)(2), (d)(1)(iv), & (e)(1)]

h. Proper eye protective equipment to prevent exposure of personnel shall be provided. [1926.353(e)(2)]

71. Wire Ropes, Chains, Hooks, and so on

a. Wire ropes, chains, ropes, and other rigging equipment shall be inspected prior to use and as necessary during use to assure their safety. Defective gear shall be removed from service. [1926.251(a)(1)]

b. Job or shop hooks and links, or makeshift fasteners, formed from bolts, rods, and so on, or other such attachments, shall not be used. [1926.251(b)(3)]

c. When U-bolts are used for eye splices, the U-bolt shall be applied so the "U" section is in contact with the dead end of the rope. [1926.251(c)(5)(i)]

d. When U-bolt wire rope clips are used to form eyes, table 16.4 shall be used to determine the number and spacing of clips.

Table 16.4 (OSHA Table H–20)

Number and Spacing of U-Bolt Wire Rope Clips			
Improved Plow Steel, Rope (Diameter in Inches)	Number of Clips		Minimum Spacing in Inches
	Drop Forged	Other Material	
1/2	3	4	3
5/8	3	4	3 3/4
3/4	4	5	4 1/2
7/8	4	5	5 1/4
1	5	6	6
1 1/8	6	6	6 3/4
1 1/4	6	7	7 1/2
1 3/8	7	7	8 1/4
1 1/2	7	8	9

72. Woodworking Machinery

a. All fixed power-driven woodworking tools shall be provided with a disconnect switch that can be either locked or tagged in the off position. [1926.304(a)]

b. All woodworking tools and machinery shall meet applicable requirements of ANSI 01.1-1961, "Safety Code for Woodworking Machinery." [1926.304(f)]

Index